# SIMPLY

# NUTRITION

## FOR COMPLETE BEGINNERS

## DK LONDON

**Project Editor** Daniel Byrne
**Project Art Editor** Daksheeta Pattni
**Editors** Michael Clark, Vicki Murrell,
Rachel Warren Chadd
**Managing Editor** Gareth Jones
**Senior Managing Art Editor** Lee Griffiths
**Production Editor** Robert Dunn
**Producer** Nancy-Jane Maun
**Jacket Design Development Manager** Sophia M.T.T.
**Jacket Designer** Akiko Kato
**Associate Publishing Director** Liz Wheeler
**Art Director** Karen Self
**Publishing Director** Jonathan Metcalf

First published in Great Britain in 2023 by
Dorling Kindersley Limited
DK, One Embassy Gardens, 8 Viaduct Gardens,
London, SW11 7BW

The authorised representative in the EEA is
Dorling Kindersley Verlag GmbH. Arnulfstr. 124,
80636 Munich, Germany

Copyright © 2023 Dorling Kindersley Limited
A Penguin Random House Company
10 9 8 7 6 5 4 3
005–334440–Nov/2023

A CIP catalogue record for this book
is available from the British Library.
ISBN: 978-0-2416-0730-5

Printed and bound in China

## For the curious
**www.dk.com**

## CONSULTANT

**Juliette Kellow** is a registered dietitian with a passion for food, nutrition, and health. A member of the British Dietetic Association, she has worked in the NHS, for the food industry, and is the former editor of *Top Santé* magazine. Juliette now works as a nutrition consultant for many food brands, industry and trade bodies, and PR agencies. She regularly writes for magazines, newspapers, and online, and is co-author of several nutrition books.

## CONTRIBUTOR

**Fiona Hunter** is a highly respected and experienced nutritionist known for her practical evidence-based approach to nutrition. After working as a dietitian in the NHS, she joined *Good Housekeeping* magazine. Fiona now works as a freelance consultant and writes for many publications, including national newspapers and magazines. She says her goal is to provide honest, unbiased advice on nutrition designed to help people make informed choices about the food they eat.

# CONTENTS

## DIGESTING AND ABSORBING FOOD

# HEALTHY
# FOOD CHOICES

# AGES
## AND STAGES

# TOP-TO-TOE
## HEALTH

# EATING HABITS
## AND DIETS

# DIET AND HEALTH CONDITIONS

# WHAT IS NUTRITION?

"If we could give every individual the right amount of nourishment and exercise, we would have found the safest way to health." This hypothesis, attributed to the ancient Greek physician Hippocrates, is confirmed by modern research which has also uncovered how nutrition works. Yet, despite all the advice now available – from doctors, fitness experts, celebrities, and food marketeers – opinions and sales tactics can make it difficult to separate science fact from science fiction.

The basics, however, are incontestable. We need protein, fat, and carbohydrates, together with fibre, in varying quantities for growth, development, and body maintenance. Vitamins, minerals, and phytonutrients in plants also play a vital role in keeping us healthy. Fluid, too, is essential – without it humans cannot survive for more than a few days. Our miraculous digestive system ensures that the body absorbs nutrients from food while packaging the rest as waste matter. Enzymes, bile, and the trillions of microbes that make up the gut microbiome perform this complex process.

A balanced and varied diet, with an emphasis on plant-based foods, can supply all the nutrients required. No single nutrient is a panacea but, within minutes, good nutrition can boost energy and improve concentration. Within days it can ease digestive problems, lift mood, and aid sleep. After a few months, it can make a difference to skin, body weight, blood pressure, and blood cholesterol. In the long term it can help protect against disorders, such as cardiovascular disease, type 2 diabetes, osteoporosis, and many cancers.

Experts estimate that, globally, eating more healthily could prevent one in five deaths and help protect our future planet. Put simply, nutrition knowledge is the first step on the road towards a healthier, longer, and more sustainable life.

# WHAT
## IS
# FOOD?

**We need to eat to stay alive.** Together with water, food is fundamental for life. It supplies the energy our organs need to function effectively, adds fluid to keep us hydrated, and all its components – protein, fat, carbohydrates, fibre, vitamins, and minerals – work together to provide life support. It is the amazing interplay of these nutrients that keeps us healthy and well.

# THE ROLE
# OF FOOD

Food is essential for life. It is defined as any solid or liquid that supplies the body with the basic materials it needs for energy production, growth, and repair. The components of food with these functions – known as nutrients – are protein, fat, carbohydrates, vitamins, and minerals. Plant foods also supply phytonutrients and dietary fibre, which are not classed as essential nutrients, although they support good health. Water is a basic requirement for life and is supplied by food and drink.

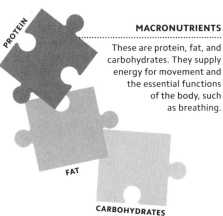

**PROTEIN**

**FAT**

**CARBOHYDRATES**

**MACRONUTRIENTS**

These are protein, fat, and carbohydrates. They supply energy for movement and the essential functions of the body, such as breathing.

**MICRONUTRIENTS**

Vitamins and minerals are needed in small amounts but are still essential for the body to develop and function (see p.22).

**VITAMINS**

**MINERALS**

**WATER**

Around 60 per cent of the adult body is made up of water. It is vital for most functions in the body (see p.42).

**PHYTONUTRIENTS**

Also known as phytochemicals, these chemicals, naturally made by plants, have been linked to many health benefits (see pp.40–41).

**DIETARY FIBRE**

Dietary fibre is a type of carbohydrate that cannot be digested (see p.18). It is essential for a healthy gut (see p.106).

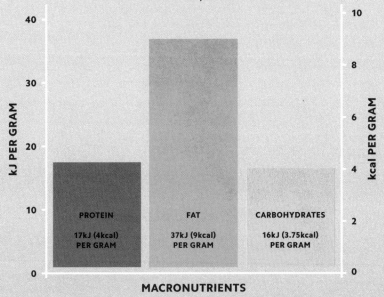

**Calculating calories**
The energy value of a food depends on its fat, protein, and carbohydrate content. As fat contains more than double the calories of protein or carbohydrate per gram, foods that contain more fat usually have more calories.

kJ PER GRAM

kcal PER GRAM

**PROTEIN**
17kJ (4kcal)
PER GRAM

**FAT**
37kJ (9kcal)
PER GRAM

**CARBOHYDRATES**
16kJ (3.75kcal)
PER GRAM

MACRONUTRIENTS

# ENERGY FROM FOOD

The energy in food is measured in kilojoules (kJ). However, many countries continue to use kilocalories (kcal) – an older energy measurement often shortened to "calories". The three main energy providers in the diet are protein, fat, and carbohydrates. Fermentation of some types of dietary fibre by gut bacteria in the large intestine produces short chain fatty acids (see p.52), which the body uses for energy, too. However, at 8kJ (2kcal) per gram, dietary fibre makes a small contribution to overall energy intake. Alcohol also supplies energy, providing 29kJ (7kcal) per gram.

# ENERGY IN VS ENERGY OUT

The amount of energy (or calories) a person requires depends on many factors (e.g. age, sex, activity levels, and body size and composition). Meeting the body's energy needs depends on balancing energy intake and energy expenditure. When this happens, body weight is maintained. However, when a diet provides more energy than is expended (positive energy balance), the extra energy is used to make adipose (fat) tissue, resulting in weight gain. When energy intake falls below expenditure (negative energy balance), stored adipose tissue is used to meet energy needs, causing weight loss. If fat stores are used up or energy intakes are low, for example when trying to lose weight, protein in muscles is broken down to supply energy.

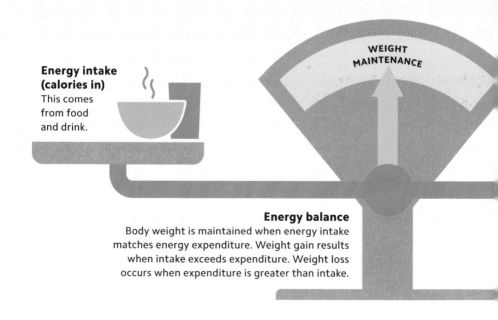

**WEIGHT MAINTENANCE**

**Energy intake (calories in)**
This comes from food and drink.

**Energy balance**
Body weight is maintained when energy intake matches energy expenditure. Weight gain results when intake exceeds expenditure. Weight loss occurs when expenditure is greater than intake.

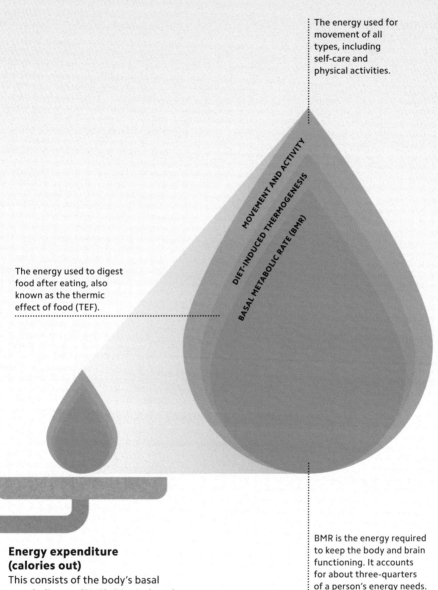

The energy used for movement of all types, including self-care and physical activities.

MOVEMENT AND ACTIVITY

DIET-INDUCED THERMOGENESIS

BASAL METABOLIC RATE (BMR)

The energy used to digest food after eating, also known as the thermic effect of food (TEF).

**Energy expenditure (calories out)**

This consists of the body's basal metabolic rate (BMR), Diet-induced thermogenesis (DIT), and all movement and activity.

BMR is the energy required to keep the body and brain functioning. It accounts for about three-quarters of a person's energy needs. BMR is influenced by body composition, sex, genes, and age, among other factors.

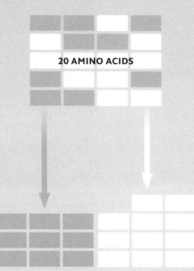

**20 AMINO ACIDS**

**Essential amino acids**
Diet must supply these. Some foods (e.g. meat, poultry, fish, eggs, dairy, soya, quinoa, buckwheat) contain the full set of nine. Other grains, pulses, nuts, and seeds contain some but not the full set.

**Non-essential amino acids**
These can be synthesized in the body so do not have to be derived from food. However, animal and plant proteins are rich sources of non-essential amino acids, which support health.

# BUILDING BLOCKS

Proteins are made from building blocks called amino acids. Humans need 20 different amino acids in order to function correctly. Nine cannot be made in the body so must come from food. These are called "essential" or "indispensable" amino acids. The remaining 11 can be made by the body (or provided by food), so are "non-essential" or "dispensable" amino acids. However, some of these become essential in certain circumstances, such as pregnancy or illness, or for children, because their bodies cannot make enough to meet their needs. These are called "conditional" amino acids.

# THE ROLE OF PROTEIN

There are thousands of different protein types and each has a specific role in the body – it is the unique combination and sequence of amino acids that determines its function. Proteins are present in every cell of the body. They are needed for the growth, repair, and maintenance of body tissues and are continually being renewed and replaced in a process called protein turnover. Some proteins have a structural role, such as in the skin, hair, nails, muscles, and bones. Others are used to make enzymes, hormones, or antibodies. Proteins can also be used to supply energy.

**Immune support**
Antibodies are proteins that protect the body from harmful bacteria and viruses.

**Energy providers**
Proteins supply energy to the body when fat and carbohydrates are in short supply.

**Muscle development**
Proteins are vital for building and maintaining muscle mass.

**Hormone function**
Many hormones, such as insulin and human growth hormone, are made from proteins.

**Enzyme activity**
Enzymes are proteins that support thousands of chemical reactions, such as those involved in digestion.

**Molecule transporters**
Many proteins act as carriers (e.g. haemoglobin is a protein that takes oxygen from the lungs to cells).

**Genetic expression**
Genes provide instructions for making specific proteins that build the body (e.g. determining eye colour and height).

**Structural support**
Some proteins give strength, rigidity, or elasticity to cells and tissues (e.g. keratin in hair and nails, and collagen in bones, tendons, and skin).

All fatty acid chains have a group of carbon, oxygen, and hydrogen atoms at one end.

**Saturated fatty acids**
All carbon atoms are linked to hydrogen atoms by single bonds. Examples include palmitic acid and stearic acid in butter.

**Trans fatty acids**
One or more double bonds have a hydrogen atom on opposite sides. This type of fatty acid is mainly formed during food processing.

**Monounsaturated fatty acids**
One double bond links two carbon atoms, attached to two adjacent hydrogen atoms. Oleic acid in olive oil is an example.

**Polyunsaturated fatty acids**
These have two or more double bonds. Examples include linoleic acid in vegetable oils and alpha-linolenic acid in walnuts.

● Oxygen ○ Hydrogen ● Carbon  Double bond

# THE STRUCTURE OF FAT

Fats are compounds, known as triglycerides, which are each made up of three fatty acids attached to a unit of glycerol. Fatty acids are termed saturated, monounsaturated, or polyunsaturated according to the structure of their carbon and hydrogen molecules (see above). Solid fats generally contain more saturated fatty acids, which can contribute to high blood cholesterol levels; most soft or liquid fats contain more unsaturated fatty acids that are essential for good health. Trans fatty acids are unsaturated, but their structure means they behave more like saturated fat, so can raise blood cholesterol.

# THE ROLE OF FAT

Fat provides the body with a concentrated source of energy. When energy needs are met, any excess is stored in fat (adipose) tissue, which protects and cushions major organs, insulates the body, and provides an energy reserve to draw on when requirements increase or food intake is reduced. Fatty acids are essential to all body cells, and while the brain contains a lot of water, fat makes up 60 per cent of its dry weight. Fats carry the fat-soluble vitamins A, D, E, and K around the body, and also supply linoleic acid and alpha-linolenic acid, essential fats the body needs for growth but cannot make itself.

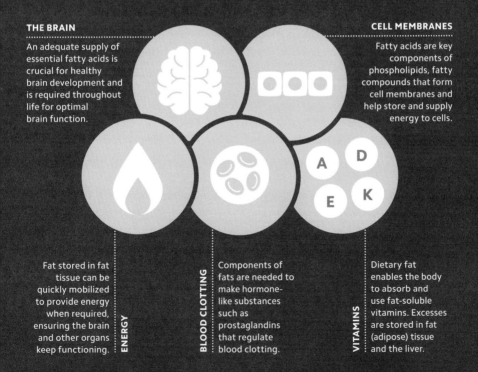

**THE BRAIN**

An adequate supply of essential fatty acids is crucial for healthy brain development and is required throughout life for optimal brain function.

**CELL MEMBRANES**

Fatty acids are key components of phospholipids, fatty compounds that form cell membranes and help store and supply energy to cells.

**ENERGY**

Fat stored in fat tissue can be quickly mobilized to provide energy when required, ensuring the brain and other organs keep functioning.

**BLOOD CLOTTING**

Components of fats are needed to make hormone-like substances such as prostaglandins that regulate blood clotting.

**VITAMINS**

Dietary fat enables the body to absorb and use fat-soluble vitamins. Excesses are stored in fat (adipose) tissue and the liver.

# SIMPLE AND COMPLEX CARBS

Carbohydrates are produced in a plant's leaves during photosynthesis, most often starting with a single molecule of glucose (sugar). Carbohydrates are classifed according to the length of the chain that is formed when glucose molecules join together chemically. Those with one or two molecules or units (see right) are described as simple carbohydrates; those with three or more are called complex carbohydrates. The latter can be split into two further categories: oligosaccharides and polysaccharides. The length of the chain determines how quickly the carbohydrate is broken down and digested. Dietary fibre is a complex carbohydrate that cannot be broken down into sugar molecules in the small intestine. It plays an important role in gut health (see p.106).

**Sugar chains**

Simple carbohydrates (indicated in white) are the building blocks for larger, more complex carbohydrates (indicated in yellow).

SIMPLE

COMPLEX

**SINGLE UNIT**

Monosaccharides are often called simple sugars. There are three types.

**GLUCOSE**
e.g. honey

**FRUCTOSE**
e.g. fruit

**GALACTOSE**
e.g. dairy

**DOUBLE UNITS**

Also known as disaccharides, these form when two monosaccharide units join. There are three types.

**MALTOSE**
e.g. molasses, beer

**SUCROSE**
e.g. sugar beet

**LACTOSE**
e.g. dairy

**MULTIPLE UNITS**

Oligosaccharides are between three and 10 molecules long. Polysaccharides contain more than 10, and often thousands.

**STARCH**
e.g. rice, potato

**GLYCOGEN**
stored glucose in the body

**CELLULOSE**
plant cell walls

**Instant energy**

Glucose travels in the blood to cells. As blood glucose levels rise, insulin is released, enabling glucose to enter and refuel cells. Insulin then helps store any excess glucose in the liver and muscles, or as fat if these stores are full.

**Liver storage**

Glucose is stored as glycogen in the liver and quickly converted back to glucose when energy is needed. During fasting, liver glycogen will last 12–18 hours, after which the body will use energy from fat and protein stores.

**Muscle storage**

Muscles also store glucose as glycogen. This can be accessed quickly if glucose is suddenly needed to supply energy – for example, during exercise.

**Fat storage**

When cells are fully fuelled and liver and muscle stores of glycogen are full, any extra glucose is used to make fatty acids, which are stored as body fat.

# THE ROLE OF CARBOHYDRATES

The main role of carbohydrates are to supply the body with energy. During digestion, carbohydrates are broken down into glucose, which is used to fuel the body's cells and organs. The brain demands the most energy and needs a continuous glucose supply for its billions of nerve cells. The body also stores small amounts of glucose in the liver and muscles as glycogen (many connected glucose molecules). Any surplus glucose is converted into fatty acids, which are stored in adipose tissue (body fat). Dietary fibre cannot be broken down into glucose molecules, but it helps regulate the body's use of glucose, keeping hunger and blood sugar in check.

# WHAT IS FIBRE?

Dietary fibre is an umbrella term for the diverse group of indigestible carbohydrates in plant foods. They all progress to the large intestine, where they are broken down by "good" gut bacteria. Prebiotics are types of fibre that stimulate the growth of these bacteria. Until recently, fibre was often termed "soluble" or "insoluble", but now it is classed according to its properties. These include its viscosity (whether it forms a gel in the gut), which affects nutrient absorption, and its fermentability, which affects gut bacteria. To benefit from the many different types of fibre, it is best to eat a wide variety of different plants. Like fibre, resistant starch is not digested in the small intestine and provides food for bacteria in the large intestine, conferring many similar health benefits.

**SOURCES OF FIBRE AND RESISTANT STARCH**

**Beta glucans**
e.g. oats and barley

**Cellulose and hemicellulose**
e.g. cereal grains, fruit, vegetables, and nuts

**Gums and mucilages**
e.g. seeds and seaweed extracts

**Lignan**
e.g. outer cereal layers and celery

**Oligosaccharides and inulin**
e.g. onions, leeks, asparagus, legumes, and chicory

**Pectins**
e.g. fruit, vegetables, nuts, legumes, and potatoes

**Resistant starch**
e.g. cooked and chilled potatoes and pasta, cereal grains, and legumes

**Lowers blood cholesterol**
Certain types of fibre bind with cholesterol and inhibit its absorption. Keeping cholesterol levels within a healthy range can lower the risk of cardiovascular disease and stroke.

**Improves glucose control**
Fibre slows glucose absorption and good intakes are linked to a lower risk of type 2 diabetes.

**Possibly aids satiety**
Improved satiety helps to curb hunger. This can potentially help to prevent weight gain that can lead to obesity and related health problems.

**Supports gut health**
Bacterial fermentation of fibre results in the production of short chain fatty acids, which may help protect against bowel cancer.

**Health benefits**
Eating plenty of fibre and resistant starch from a diverse range of plant sources provides the body with the benefits of each type of fibre, offering protection against various diseases.

**Reduces gut transit time and gives bulky, soft stools**
This helps to prevent constipation, diverticular disease, and bowel cancer.

## Vitamins

There are 13 essential vitamins and they are either water-soluble or fat-soluble, which affects how they are stored in the body.

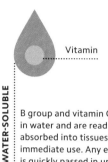

Vitamin

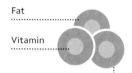

Fat

Vitamin

**WATER-SOLUBLE**

B group and vitamin C dissolve in water and are readily absorbed into tissues for immediate use. Any excess is quickly passed in urine.

**FAT-SOLUBLE**

Vitamins A, D, E, and K are absorbed more easily in the presence of dietary fat and stored in the body in the liver and fatty tissue for future use.

# VITAMINS AND MINERALS

Vitamins and minerals are essential for our bodies to develop and function normally. Collectively, they are known as micronutrients as they are needed in small, often tiny, amounts. A healthy balanced diet should provide the full range of micronutrients needed to meet all requirements and prevent deficiences (which can lead to serious health issues), although there are times in life when supplements may be necessary (see p.83). Supplement use should always follow health professional advice or packaging dosage guidance.

## Minerals

These are divided into two groups – major minerals (or macrominerals) and trace elements, depending on the amount needed to support health.

**MAJOR MINERALS**

Needed in larger amounts, these include calcium, phosphorus, potassium, sodium, chlorine, magnesium, and sulphur.

**TRACE ELEMENTS**

Needed in tiny amounts, these include iron, zinc, copper, manganese, selenium, iodine, fluoride, chromium, and molybdenum.

# THE VISION HELPER

There are two forms of vitamin A: retinol, which is found in animal foods, and carotenoids (e.g. beta-carotene), which are found in leafy greens and orange, red, and yellow fruit and vegetables. Retinol is also called "preformed" vitamin A as it is the active form of the vitamin, whereas carotenoids (also called provitamin A) have to be converted to retinol before being used by the body. This nutrient is essential for good vision (especially in dim light), plus it's vital for immune function, healthy skin and maintaining mucous membranes (e.g. the lining of the nose, mouth, lungs, and stomach). Vitamin A deficiency is rare in the UK but is a serious health issue in developing countries and can lead to increased susceptibility to infections, infertility, delayed growth, skin problems, and night blindness. In contrast, excessive amounts can potentially be harmful when trying to conceive and during pregnancy (see p.94).

## KEY SOURCES

**FULL-FAT DAIRY PRODUCTS (RETINOL)**

**EGGS (RETINOL)**

**ORANGE/RED/ YELLOW FRUIT AND VEGETABLES (BETA-CAROTENE)**

**LEAFY GREENS (BETA-CAROTENE)**

## KEY FUNCTIONS

**BENEFITS VISION**

**SUPPORTS THE IMMUNE SYSTEM**

**PROMOTES HEALTHY SKIN**

**MAINTAINS MUCOUS MEMBRANES**

# THE MULTI-TASKERS

There is no single vitamin B. Instead, several vitamins with similar properties and functions form the B group. Their main roles are to convert food into energy, support nerve and brain function, and make red blood cells. B-group vitamins are found in many foods (see examples in the table), so eating a varied diet usually fulfils needs. However, as B12 is only found naturally in animal products, people following a vegan diet may need a supplement. Health authorities in many countries (e.g. the UK, US, Canada, and Australia) also advise that those trying to get pregnant and in the first trimester of pregnancy take a folic acid supplement (the synthetic version of folate; vitamin B9) to help prevent neural tube defects, such as spina bifida, in unborn babies.

| | |
|---|---|
| **VITAMIN B1 (THIAMINE)** | Red meat, wholegrains, fortified cereals, bread, pulses, nuts, seeds |
| **VITAMIN B2 (RIBOFLAVIN)** | Red meat, poultry, oily fish, eggs, dairy, fortified cereals, mushrooms, mycoprotein |
| **VITAMIN B3 (NIACIN)** | Red meat, poultry, white and oily fish, shellfish, wholegrains, fortified cereals, bread, soya, nuts, seeds |
| **VITAMIN B5 (PANTOTHENIC ACID)** | Red meat, poultry, shellfish, eggs, dairy, fortified cereals, nuts, seeds, mushrooms |
| **VITAMIN B6 (PYRIDOXINE)** | Red meat, poultry, white and oily fish, fortified cereals, soya, nuts, seeds |
| **VITAMIN B7 (BIOTIN)** | White fish, eggs, soya, nuts, seeds, mushrooms, mycoprotein |
| **VITAMIN B9 (FOLATE)** | Leafy vegetables, eggs, fortified cereals, nuts, seeds, mycoprotein |
| **VITAMIN B12 (COBALAMIN)** | Red meat, poultry, white and oily fish, shellfish, eggs, dairy, fortified cereals |

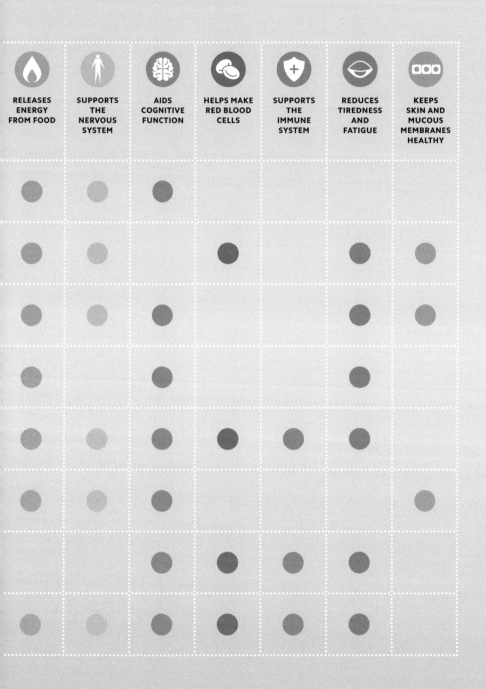

## KEY SOURCES

**CITRUS FRUIT**

**PEPPER**

**BERRIES**

**LEAFY GREENS**

THE IMMUNE SUPPORTER

Vitamin C helps the immune system fight infection and is essential for the production of collagen – a protein that heals wounds and gives structure and strength to skin, bones, cartilage, blood vessels, teeth, and gums. Vitamin C supports the nervous system, has been linked to psychological wellbeing and, as an antioxidant, helps protect cells from damage. It also helps the body absorb the type of iron found in plant foods, making it a useful nutrient for plant-based diets (see p.124). A severe deficiency is rare but can lead to a disease called scurvy. Large doses are unlikely to be harmful but can be a risk for kidney stones.

## KEY FUNCTIONS

**SUPPORTS THE IMMUNE SYSTEM**

**SUPPORTS THE NERVOUS SYSTEM**

**PROVIDES ANTIOXIDANT SUPPORT**

**AIDS IRON ABSORPTION**

## KEY SOURCES

**OILY FISH**

**EGGS**

**FORTIFIED CEREALS**

**UV-ENRICHED MUSHROOMS**

# THE SUNSHINE VITAMIN

Vitamin D is also known as the sunshine vitamin because, although present in some foods, we get most of it when our skin is exposed to direct sunlight. Vitamin D helps the body absorb calcium and phosphorus from food, which are integral to bone development during childhood (when a deficiency can cause rickets) and bone strength in adulthood (when a deficiency can cause osteoporosis). It is also essential for regulating inflammation and immune function and supporting muscle function. Vitamin D deficiency is one of the most common nutritional deficiencies in the world and health authorities often recommend supplement use during winter months. However, too much vitamin D can weaken bones and damage the kidneys and heart.

## KEY FUNCTIONS

**KEEPS BONES HEALTHY**

**KEEPS TEETH HEALTHY**

**SUPPORTS THE IMMUNE SYSTEM**

**MAINTAINS MUSCLE FUNCTION**

Vitamin E is found in the membrane of cells, where it acts as an antioxidant, helping to neutralize harmful molecules called free radicals that can damage cells. Vitamin E also supports the immune system, and some research suggests that good intakes may protect against heart disease, Alzheimer's disease, and age-related macular degeneration. However, results are mixed for the vitamin in supplement form. As it is a fat-soluble vitamin, the body can store it for future use; a deficiency, although rare, may weaken immunity. Too much vitamin E in supplement form may harm health, for example increasing the risk of bleeding.

## THE CELL DEFENDER

### KEY SOURCES

**VEGETABLE OIL**

**SEEDS**

**NUTS**

**AVOCADO**

### KEY FUNCTIONS

**PROVIDES ANTIOXIDANT SUPPORT**

**SUPPORTS THE IMMUNE SYSTEM**

**SUPPORTS THE NERVOUS SYSTEM**

**BENEFITS VISION**

# THE CLOTTING NUTRIENT

Vitamin K is essential for blood clotting. As babies are born with a vitamin K deficiency, they are given an injection of it shortly after birth. Vitamin K has a role in bone and joint health, too, as it supports the efficient functioning of various proteins in bone and cartilage. It may also prevent calcification in the arteries, keeping blood vessels healthy. The main form of vitamin K is phylloquinone or vitamin K1, found in leafy greens and soya. The other form, menaquinones or vitamin K2, is produced by bacteria in the gut and is present in fermented foods and animal products. Vitamin K is broken down in the body quickly so rarely reaches harmful levels and, as a fat-soluble vitamin, a deficiency is rare.

## KEY SOURCES

**LEAFY GREENS (K1)**

**SOYA (K1)**

**EGGS (K2)**

**FERMENTED FOODS (K2)**

## KEY FUNCTIONS

**ENABLES BLOOD CLOTTING**

**KEEPS BONES HEALTHY**

**KEEPS JOINTS HEALTHY**

**KEEPS BLOOD VESSELS HEALTHY**

## KEY SOURCES

**DRIED FRUIT**  **AVOCADO**  **POTATOES**  **BANANAS**

## KEY FUNCTIONS

**REGULATES FLUID LEVELS**  **REGULATES BLOOD PRESSURE**  **SUPPORTS THE NERVOUS SYSTEM**  **MAINTAINS MUSCLE FUNCTION**

# THE FLUID REGULATOR

Potassium works closely with sodium to maintain the balance of fluids in the body. While sodium controls fluid levels outside of the cells, potassium helps to regulate them within cells. Potassium also relaxes blood vessel walls, which is why it is essential for controlling blood pressure and why potassium-rich foods are a key part of the DASH diet (see p.137). Potassium is also needed for normal nerve function and muscle contraction. It is sometimes referred to as an electrolyte because, when dissolved in water, it creates a small electrical charge that activates various cell and nerve functions (e.g. maintaining fluid balance and muscle contraction). If levels become too low, or high, muscle weakness and arrhythmias (abnormal heart rhythms) may develop.

## KEY SOURCES

**DAIRY PRODUCTS**

**TOFU**

**FORTIFIED PLANT MILKS**

**BREAD**

. . . . . . . . . . . . . . . . . . . . . . . .

## KEY FUNCTIONS

**KEEPS BONES HEALTHY**

**KEEPS TEETH HEALTHY**

**AIDS BLOOD CLOTTING**

**MAINTAINS MUSCLE FUNCTION**

# THE BONE BUILDER

Calcium is vital for strong teeth and bones (the latter is where over 98 per cent of body calcium is stored). It is also needed for blood clotting, muscle function, supporting digestive enzymes, and nerve transmission. If the diet lacks calcium, the body tops up blood levels by drawing calcium out of the bones. If bones lack calcium, they can weaken, causing a condition called osteoporosis (see p.150). The calcium in dairy products is more easily absorbed by the body than that in plant foods (see p.68). Pulses, nuts, seeds, dried fruit, and spinach contain calcium but also oxalates and/or phytates that inhibit calcium absorption. However, plant-based diets can still fulfil calcium needs, especially if calcium-enriched foods are eaten.

## KEY SOURCES

**DAIRY PRODUCTS**

**FISH**

**SEEDS**

**WHOLEGRAINS**

## KEY FUNCTIONS

**KEEPS BONES HEALTHY**

**KEEPS TEETH HEALTHY**

**ENABLES CELL MEMBRANE FUNCTION**

**RELEASES ENERGY FROM FOOD**

# THE BONE HELPER

Phosphorus works together with calcium to keep bones and teeth healthy. It is the second most abundant mineral in the body after calcium, with around 85 per cent stored in the skeleton and teeth. Phosphorus also plays a role in providing energy for reactions in every cell of the body; the production, storage, and transfer of energy depends on adequate amounts. It occurs naturally in many foods but also comes from processed foods in the form of phosphates (added as preservatives, colourings, or emulsifiers), so most people obtain enough from their diet. Like calcium, phosphorus is more easily absorbed from animal foods than plant foods. However, nuts, seeds, and wholegrains are still good sources.

# THE WORKHORSE

Magnesium's many functions in the body include maintaining the health of muscles, helping to support muscle contraction, nerve impulse transmission, and a steady heartbeat. It also keeps bones and teeth strong. Magnesium is involved in more than 300 enzyme reactions, too, including making proteins and regulating blood pressure and blood glucose. Low intakes have been linked to an increased risk of type 2 diabetes, possibly because of its role in controlling blood glucose. However, there is insufficient evidence to support taking supplements, unless the diet is deficient. Signs of a deficiency include muscle cramps, weakness, and an abnormal heart rhythm but, as magnesium is found in a wide range of foods, this is uncommon.

**KEY SOURCES**

NUTS

SEEDS

LEAFY GREENS

WHOLEGRAINS

**KEY FUNCTIONS**

KEEPS BONES HEALTHY

KEEPS TEETH HEALTHY

MAINTAINS MUSCLE FUNCTION

SUPPORTS THE NERVOUS SYSTEM

## KEY SOURCES

**RED MEAT**

**EGGS**

**FORTIFIED CEREALS**

**PULSES**

## KEY FUNCTIONS

**MAKES RED BLOOD CELLS**

**AIDS COGNITIVE FUNCTION**

**SUPPORTS THE IMMUNE SYSTEM**

**RELEASES ENERGY FROM FOOD**

# THE OXYGEN TRANSPORTER

Iron is essential for making haemoglobin, a protein in red blood cells that transports oxygen around the body. There are two different types of iron. Meat provides haem iron, which is easily absorbed by the body. The non-haem iron in eggs and plant foods is less easily absorbed, but they are still a good source and studies show that well-planned plant-based diets can easily fulfil iron needs. Combining non-meat iron-rich foods with vitamin C also aids absorption. Tea, coffee, and red wine contain tannins that inhibit the absorption of non-haem iron, so consume separately. A diet deficient in iron means fewer red blood cells are made and so less oxygen reaches organs and tissues. This is called iron-deficiency anaemia (see p.152).

# THE ALL-ROUNDER

Zinc is present in every cell in the body and is vital for cell division, immune function, fertility, DNA synthesis, and protein production. The constant renewal and repair of the skin, hair, and nails relies on zinc, especially wound healing, and the mineral is also needed for the eyes, bones, and brain to function properly. Zinc from meat, fish, and shellfish is more readily absorbed by the body than zinc from plant foods. Phytates (a natural component of plants that is especially high in wholegrains, nuts, and seeds) binds with zinc and inhibits intestinal absorption. Soaking and sprouting grains reduces the binding effects but people following a plant-based diet may also benefit from supplements.

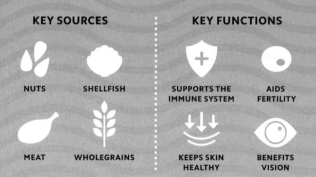

**KEY SOURCES**

NUTS

SHELLFISH

MEAT

WHOLEGRAINS

**KEY FUNCTIONS**

SUPPORTS THE IMMUNE SYSTEM

AIDS FERTILITY

KEEPS SKIN HEALTHY

BENEFITS VISION

# THE FREE RADICAL FIGHTER

Selenium is a trace mineral found in the soil and the sea, and so is present in seafood and most foods grown in the ground. Animals that eat such foods (e.g. lamb, chicken) are also a good source. However, the selenium content of food depends on levels in the soil and this varies dramatically around the world. The nutrient plays a vital role in thyroid function, which helps control metabolism, and works as an antioxidant, destroying free radicals that, if left to accumulate, can damage cells (including the DNA and protein within them) and increase the risk of heart disease and cancer. Selenium is also needed for the immune system and supports fertility through its role in sperm production.

**KEY SOURCES**

**NUTS (ESPECIALLY BRAZIL NUTS)**

**FISH**

**MEAT**

**SEEDS**

**KEY FUNCTIONS**

**REGULATES METABOLISM**

**PROVIDES ANTIOXIDANT SUPPORT**

**SUPPORTS THE IMMUNE SYSTEM**

**AIDS FERTILITY**

## KEY SOURCES

**FISH**

**SHELLFISH**

**SEAWEED**

**DAIRY PRODUCTS**

## KEY FUNCTIONS

**SUPPORTS GROWTH**

**REGULATES METABOLISM**

**AIDS COGNITIVE FUNCTION**

**SUPPORTS THE NERVOUS SYSTEM**

# THE THYROID REGULATOR

Found mainly in seafood and animal products, iodine is required to make thyroid hormones that are essential for growth and metabolism. Low iodine status is increasingly prevalent, particularly in young people and pregnant women, and this is concerning as even a mild deficiency may affect a baby's brain development during pregnancy and its future mental capacity and IQ. As many iodine-rich foods are excluded from a plant-based diet (see p.124), extra care is needed to guard against a deficiency. Fish and dairy products are good sources. Seaweed contains a high concentration of iodine but should be eaten no more than once a week, especially during pregnancy, as excessively high amounts can disturb thyroid function.

# BEST OF THE REST

Alongside iron, zinc, selenium, and iodine, several other trace elements are needed for the body to stay healthy. Some of these (e.g. copper, manganese, and molybdenum) are essential. Others (e.g. chromium and fluoride) are recommended for their health benefits. Although there is insufficient evidence to provide specific guidelines on the amounts needed for different ages or sexes, many countries set ranges for suitable amounts, or safe and adequate intakes. Maximum upper limits are also given as many trace elements are toxic in high amounts.

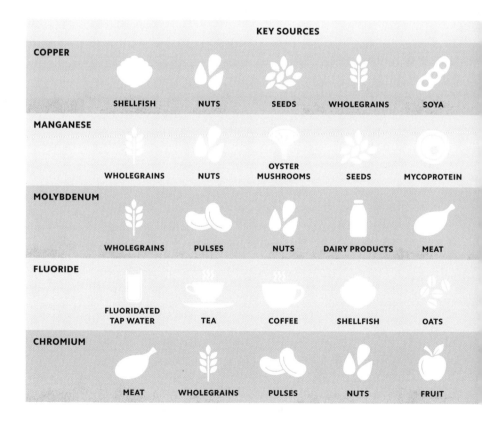

**KEY SOURCES**

**COPPER**
SHELLFISH | NUTS | SEEDS | WHOLEGRAINS | SOYA

**MANGANESE**
WHOLEGRAINS | NUTS | OYSTER MUSHROOMS | SEEDS | MYCOPROTEIN

**MOLYBDENUM**
WHOLEGRAINS | PULSES | NUTS | DAIRY PRODUCTS | MEAT

**FLUORIDE**
FLUORIDATED TAP WATER | TEA | COFFEE | SHELLFISH | OATS

**CHROMIUM**
MEAT | WHOLEGRAINS | PULSES | NUTS | FRUIT

> Deficiencies of trace elements are extremely rare, as the body only requires a small amount of each trace element for good health.

| KEY FUNCTIONS | NEED TO KNOW |
|---|---|
| Vital to the immune and nervous systems, copper releases energy from food and moves iron around the body. It is also an antioxidant and keeps connective tissue healthy. | Although rare, high zinc intake (e.g. from long-term overuse of supplements) can interfere with copper absorption, making copper deficiency more likely. |
| Manganese is important for bone health and for making connective tissue. It also helps to release energy from food and acts as an antioxidant. | The manganese-rich germs of grains are removed during refining (see p.67), so refined foods have around three times less manganese than wholegrain foods. |
| This trace element is needed for enzymes involved in DNA and protein metabolism. Molybdenum also helps to break down toxic substances and drugs in the body. | Molybdenum is found in a wide range of foods. The quantity can vary based on the amount of molybdenum in the soil and water where the food is grown. |
| Fluoride is important for tooth remineralization, which strengthens the enamel and protects teeth from decay. Fluoride also helps keep bones strong. | Fluoride is often added to toothpaste and tap water. In excessive amounts, it can cause dental fluorosis – in most cases, small spots or lines – in children's teeth. |
| Chromium helps to metabolize protein, fat, and carbohydrates in food, and is needed for insulin function. This means it helps to maintain blood glucose levels. | Although some studies suggest chromium may help combat insulin resistance and type 2 diabetes, there is not enough evidence to prove this connection. |

# EATING THE RAINBOW

**STRAWBERRIES**

**TOMATOES**

Phytonutrients (also called phytochemicals) are chemicals naturally made in plants. They are often concentrated in the outer layers, where they help protect against disease, pests, and UV light. In fruit and vegetables, they are often responsible for the flavours, smells, and colours – such as the bitterness of leafy vegetables, the pungent aroma of garlic, and the many vibrant reds, purples, oranges, and greens. Research shows that, when consumed, phytonutrients may have human health benefits. For example, many act as antioxidants and so help to protect cells from damage caused by free radicals (unstable molecules that are a byproduct of metabolism). As plant colour is an indicator of nutrient content, eating a rainbow of brightly coloured plant-based foods ensures a good mix.

**CARROTS**

**SWEETCORN**

**AVOCADO**

**LEAFY GREENS**

**RED CABBAGE**

**BLUEBERRIES**

> There are thousands of phytonutrients and scientific research is just beginning to understand all their potential.

**POTATOES**

**MUSHROOMS**

## Red

Some red fruits and vegetables contain lycopene, which may protect against some cancers and reduce blood cholesterol; anthocyanins (see below); and ellagic acid, shown to help fight inflammation and slow cancer growth.

WATERMELON   CHERRIES

## Orange/yellow

This group includes alpha- and beta-carotene and beta-cryptoxanthin. These carotenoids make vitamin A, which is important for skin, vision, and immune function, and may support heart health and protect against some cancers.

APRICOTS   SQUASH

## Green

Many green fruits and vegetables contain lutein and zeaxanthin, which support vision and protect against age-related macular degeneration; and glucosinolates, which may help protect against and also inhibit cancer.

PEAS   BROCCOLI

## Purple/blue

This colour represents polyphenols, such as anthocyanins, which are linked to better memory and cognition and heart health; and betacyanins, which may protect against bacteria and viruses, as well as some cancers.

GRAPES   AUBERGINE

## White/tan

This group includes quercetin, which may aid heart health and circulation and also act as an anti-inflammatory; and allicin, which may lower blood pressure and cholesterol and protect against some cancers.

ONION   GARLIC

Water is vital for most bodily functions, including transporting nutrients and oxygen in the blood, and removing waste products and toxins. It also lubricates and cushions joints, aids digestion, keeps skin healthy, and controls body temperature. Around three-quarters of the brain is made up of water, so even mild dehydration can affect mental performance and cause headaches, confusion, and tiredness. Prolonged dehydration can lead to constipation, urinary tract infections, and kidney stones. The body loses water through urine, sweat, and breathing. Hot weather, physical activity, breastfeeding, diuretic medication, fever, and menopausal hot flushes all increase fluid needs. Excessive water intake dilutes the blood and lowers sodium levels. Known as "hyponatraemia", it can cause muscle cramps, seizures, and coma.

# THE LIQUID OF LIFE

# STAYING HYDRATED

Drinking enough fluids each day is essential for good health. How much a person needs to drink depends on their age, size, activity levels, and the climate they live in. With so many factors at play, popular advice to drink 6–8 glasses of fluids a day is an approximate guide. Water is the best form of hydration but, except for alcohol, all liquids (including tea, coffee, milk, and soft drinks) count towards fluid requirements. However, it is best to limit those high in sugar. The colour of urine can indicate how well the body is hydrated. Dark-coloured urine is a warning sign of dehydration. However, red or dark brown urine may indicate blood is present so seek medical advice.

Around a fifth of the fluid in diets comes from food (e.g. a strawberry is 92 per cent water).

**HYDRATED**

1

2

3

4

5

6

7

8

**DEHYDRATED**

**COLOURS 1–3**
Urine that is pale yellow and odourless suggests a person is hydrated.

**COLOURS 4–8**
Urine that is bright or dark yellow and/or has an unpleasant smell are signs of dehydration. Other symptoms include a dry mouth, headache, fatigue, extreme thirst, confusion, dizziness, and absence of tears when crying.

# DIGES
## AND
# ABSOR
## FOOD

# TING

# BING

**The digestive process** converts food into energy and nutrients. Knowing how it works and how different organs keep the digestive system functioning efficiently helps us understand why good nutrition matters. Every cell in the body uses nutrients for essential functions such as making enzymes and hormones, building and strengthening bones, and growing and repairing tissue. The gut microbiome has itself become an important area of research, as the composition of its trillions of microbes is now thought to influence both physical and mental health.

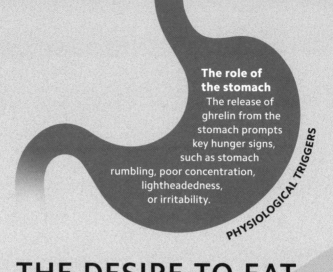

**The role of
the stomach**
The release of
ghrelin from the
stomach prompts
key hunger signs,
such as stomach
rumbling, poor concentration,
lightheadedness,
or irritability.

PHYSIOLOGICAL TRIGGERS

# THE DESIRE TO EAT

Appetite, the desire to eat or drink, can be triggered by
physiological or psychological factors. The body needs energy
and nutrients to survive, so the brain and stomach are hard-
wired to stimulate hunger when the stomach is empty. An
empty stomach produces the hormone ghrelin, which sends
signals to the brain, prompting hunger. When the stomach is
full, the body produces the hormone leptin, which turns off
hunger. Psychological factors, such as mood (see p.118), as
well as external cues, can also produce the desire to eat.

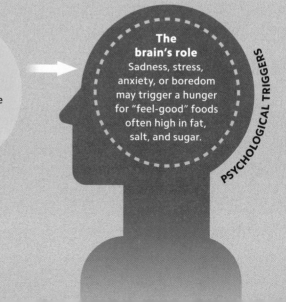

**External factors**
The sight or smell of
food can prompt the
senses to signal a desire
to eat, even when the
stomach is full.

**The
brain's role**
Sadness, stress,
anxiety, or boredom
may trigger a hunger
for "feel-good" foods
often high in fat,
salt, and sugar.

PSYCHOLOGICAL TRIGGERS

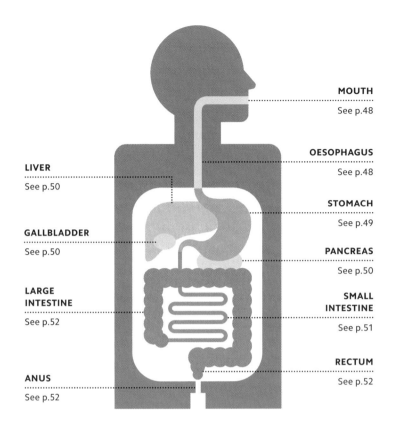

MOUTH
See p.48

OESOPHAGUS
See p.48

LIVER
See p.50

STOMACH
See p.49

GALLBLADDER
See p.50

PANCREAS
See p.50

LARGE
INTESTINE
See p.52

SMALL
INTESTINE
See p.51

RECTUM
See p.52

ANUS
See p.52

# TURNING FOOD INTO NUTRIENTS

Digestion – the conversion of food into nutrients the body can use for energy, growth, and repair – takes place in the gastrointestinal tract or alimentary canal, which extends from the mouth to the anus. The process starts in the mouth, where food is first broken down by physical chewing and the chemical action of enzymes in saliva. It continues in the stomach and small intestine, where most nutrients are absorbed into the bloodstream. The large intestine processes any undigested food, extracts water, and compacts the remains into waste, excreted via the anus.

# CHEWING AND SWALLOWING

The digestive process is stimulated before food enters the mouth. Seeing, smelling, and anticipating food encourages glands in the mouth to produce saliva, which contains digestive enzymes. As the teeth bite and chew on food, saliva moistens it and begins to break it down chemically. When a piece of chewed food (bolus) is swallowed, the epiglottis, a flap of tissue in the throat, closes to ensure the bolus goes into the oesophagus (food pipe), rather than the adjacent trachea (windpipe). The muscular oesophagus propels each bolus towards the stomach.

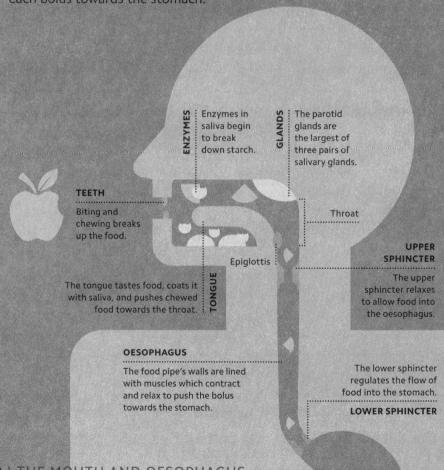

**ENZYMES**
Enzymes in saliva begin to break down starch.

**GLANDS**
The parotid glands are the largest of three pairs of salivary glands.

**TEETH**
Biting and chewing breaks up the food.

Throat

**UPPER SPHINCTER**
The upper sphincter relaxes to allow food into the oesophagus.

Epiglottis

**TONGUE**
The tongue tastes food, coats it with saliva, and pushes chewed food towards the throat.

**OESOPHAGUS**
The food pipe's walls are lined with muscles which contract and relax to push the bolus towards the stomach.

The lower sphincter regulates the flow of food into the stomach.

**LOWER SPHINCTER**

# BREAKING FOOD DOWN

**MUSCLE**
Three layers of smooth muscle in the stomach wall – extending lengthwise, laterally, and diagonally – churn food to break it down and mix it with gastric juice.

**GASTRIC JUICE**
The stomach lining secretes acidic gastric juice to break down food and kill pathogens.

**GASTRIC ENZYMES**
Pepsin, which breaks down protein, is the principal gastric enzyme.

**PYLORIC SPHINCTER**
The sphincter, a muscular valve, controls the flow of chyme into the small intestine.

**CHYME**
Mixed with gastric juice, food particles become a pulpy acidic paste called chyme.

It takes about six seconds for food to reach the stomach after leaving the mouth. It remains in the stomach for two to five hours depending on its composition – protein-rich and fatty foods take longer to digest than simple carbohydrates. During this time the stomach muscles contract and relax to churn food particles and mix them with gastric juice to further break them down. The food leaves the stomach as a semi-fluid paste called chyme.

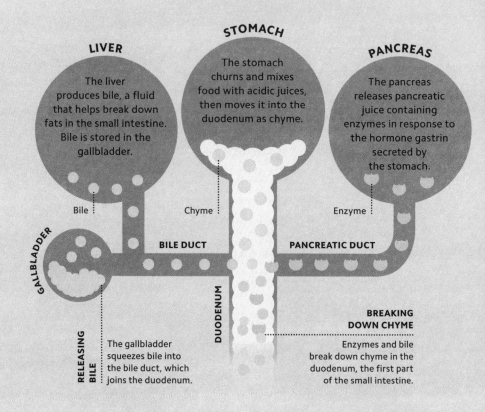

## LIVER
The liver produces bile, a fluid that helps break down fats in the small intestine. Bile is stored in the gallbladder.

## STOMACH
The stomach churns and mixes food with acidic juices, then moves it into the duodenum as chyme.

## PANCREAS
The pancreas releases pancreatic juice containing enzymes in response to the hormone gastrin secreted by the stomach.

Bile

Chyme

Enzyme

**BILE DUCT**

**PANCREATIC DUCT**

GALLBLADDER

DUODENUM

**RELEASING BILE**
The gallbladder squeezes bile into the bile duct, which joins the duodenum.

**BREAKING DOWN CHYME**
Enzymes and bile break down chyme in the duodenum, the first part of the small intestine.

# SUPPORTING DIGESTION

After leaving the stomach as chyme, food passes into the first part of the small intestine, the duodenum. Here, bile from the gallbladder and pancreatic juice from the pancreas mix with chyme to help break down protein, fat, and carbohydrates. The release of bile, which the liver produces and the gallbladder stores, is prompted by cholecystokinin (CKK), a hormone triggered when fatty foods enter the duodenum. Pancreatic juice contains enzymes – lipase to digest fat, trypsinogen to digest protein and peptides, and pancreatic amylase to digest starches, plus bicarbonate to neutralize the acidic chyme.

# ABSORBING NUTRIENTS

Most digestion and absorption of nutrients into the bloodstream takes place in the small intestine, the longest part of the digestive tract. The combined length of its three sections – the duodenum, and the coiled jejunum and ileum – is around 7m (23ft). Food remains in the small intestine for around four hours, during which bile and enzymes break it down, releasing nutrients that pass into the bloodstream. The small intestine has a rich blood supply and its inner lining is covered with thousands of villi – tiny hair-like projections through which nutrients and water can be absorbed.

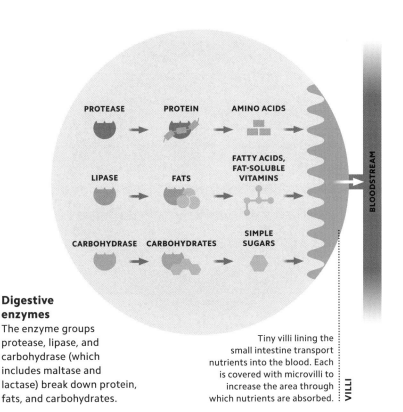

PROTEASE      PROTEIN      AMINO ACIDS

LIPASE      FATS      FATTY ACIDS, FAT-SOLUBLE VITAMINS

CARBOHYDRASE   CARBOHYDRATES   SIMPLE SUGARS

BLOODSTREAM

VILLI

**Digestive enzymes**
The enzyme groups protease, lipase, and carbohydrase (which includes maltase and lactase) break down protein, fats, and carbohydrates.

Tiny villi lining the small intestine transport nutrients into the blood. Each is covered with microvilli to increase the area through which nutrients are absorbed.

# PROCESSING WASTE

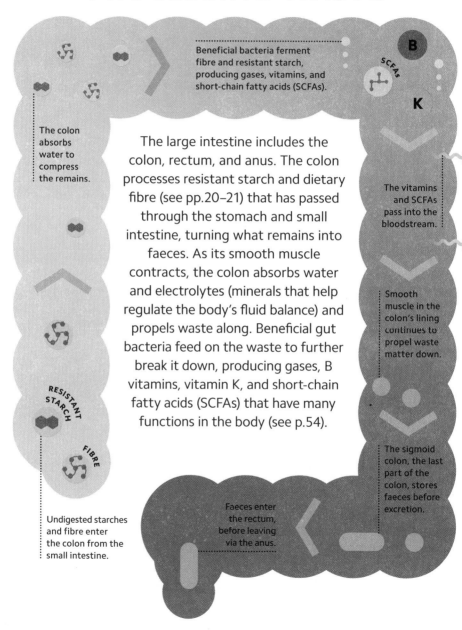

Beneficial bacteria ferment fibre and resistant starch, producing gases, vitamins, and short-chain fatty acids (SCFAs).

**B**

SCFAs

**K**

The colon absorbs water to compress the remains.

The vitamins and SCFAs pass into the bloodstream.

Smooth muscle in the colon's lining continues to propel waste matter down.

The large intestine includes the colon, rectum, and anus. The colon processes resistant starch and dietary fibre (see pp.20–21) that has passed through the stomach and small intestine, turning what remains into faeces. As its smooth muscle contracts, the colon absorbs water and electrolytes (minerals that help regulate the body's fluid balance) and propels waste along. Beneficial gut bacteria feed on the waste to further break it down, producing gases, B vitamins, vitamin K, and short-chain fatty acids (SCFAs) that have many functions in the body (see p.54).

RESISTANT STARCH

FIBRE

The sigmoid colon, the last part of the colon, stores faeces before excretion.

Undigested starches and fibre enter the colon from the small intestine.

Faeces enter the rectum, before leaving via the anus.

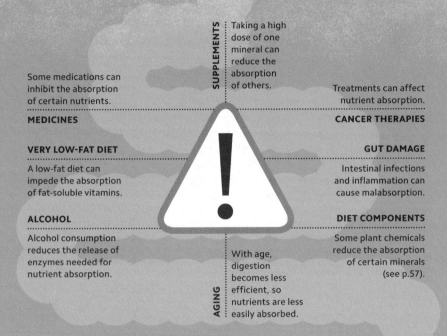

**SUPPLEMENTS**
Taking a high dose of one mineral can reduce the absorption of others.

**MEDICINES**
Some medications can inhibit the absorption of certain nutrients.

**VERY LOW-FAT DIET**
A low-fat diet can impede the absorption of fat-soluble vitamins.

**ALCOHOL**
Alcohol consumption reduces the release of enzymes needed for nutrient absorption.

**CANCER THERAPIES**
Treatments can affect nutrient absorption.

**GUT DAMAGE**
Intestinal infections and inflammation can cause malabsorption.

**DIET COMPONENTS**
Some plant chemicals reduce the absorption of certain minerals (see p.57).

**AGING**
With age, digestion becomes less efficient, so nutrients are less easily absorbed.

# NUTRIENT BLOCKERS

When food is digested in the gastrointestinal tract, it is broken down into nutrients for absorption into the bloodstream to provide energy and build healthy cells and tissues. However, diet, health problems, and medical treatments are among the factors that can reduce the body's ability to absorb nutrients effectively (see above). Malabsorption, which may result in abdominal pain, bloating, and diarrhoea, can be caused by an infection, such as gastroenteritis, or a long-term problem, such as untreated coeliac disease (see p.146). Chronic malabsorption can lead to problems such as unintentional weight loss, muscle wasting, vitamin and mineral deficiencies, and delayed growth in children.

**VITAMINS**

Beneficial gut bacteria synthesize B vitamins and vitamin K, which have many important functions.

B

K

**HARMFUL MICROBES**

"Bad" microbes, linked to allergies and bowel inflammation, are crowded out when "good" bacteria are varied and plentiful.

## A MICROBE POWERHOUSE

Around 100 trillion microbes living in the human gut, mostly in the large intestine, make up the gut microbiome. Its mix of bacteria, fungi, and viruses, which generally co-exist without causing harm, is unique to each individual. The gut microbiome keeps the gut healthy and has other far-reaching health benefits, such as supporting immunity. It also communicates with the brain – termed the gut-brain axis – via multiple pathways, and influences brain function. Factors including stress, illness, and antibiotics can upset the balance of gut bacteria. Eating fruit, vegetables, wholegrains, and fermented foods helps keep the balance stable (see opposite).

**UNDIGESTED STARCH**

Resistant starch fuels the beneficial bacteria that make short-chain fatty acids (SCFAs).

**HELPFUL ACIDS**

SCFAs' many roles include strengthening the gut lining to help ensure only nutrients pass into the bloodstream.

# A HEALTHY BALANCE

The complex range of bacteria, fungi, and viruses that make up the gut microbiome is thought to play a key role in many aspects of health. Its composition is influenced by factors such as genetics, lifestyle, age, and especially diet. Studies suggest a plant-based diet helps increase the diversity of "good" bacteria and maintains a healthy microbe balance. An imbalance (dysbiosis) may prompt digestive problems, weak immunity, heart disease, and – according to research into the gut-brain axis (see opposite) – even depression, anxiety, and obesity.

**Good gut health**
A varied, plant-based diet that also includes probiotic foods such as yogurt and sauerkraut helps to encourage a healthy balance of beneficial gut bacteria.

**Poor gut health**
A diet that lacks diversity, especially of plant-based foods, as well as smoking, stress, and lack of sleep can deplete "good" bacteria and lead to poorer gut health.

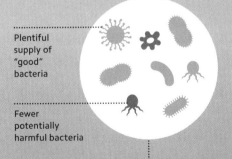

Plentiful supply of "good" bacteria

Fewer "good" bacteria

Fewer potentially harmful bacteria

More potentially harmful microbes

**A VARIED MIX**
In a healthy gut, a varied mix of microbes includes many "good" bacteria that maintain the balance and create SCFAs and some vitamins.

**UNBALANCED**
Having fewer "good" gut bacteria allows harmful microbes to proliferate, leading to digestive and other health problems.

**NON-HAEM IRON**
(e.g. fortified cereals)

**VITAMINS A AND K**
(e.g. leafy salad)

**CALCIUM**
(e.g. cheese)

**LYCOPENE**
(e.g. tomatoes)

+

**VITAMIN C**
(e.g. orange juice)

**FATS**
(e.g. olive oil)

**VITAMIN D**
(e.g. eggs)

**FATS**
(e.g. avocado)

**TO INCREASE ABSORPTION OF:**

**IRON**
Consuming vitamin C-rich foods, such as fruit and vegetables, with fortified or plant-based sources of non-haem iron boosts iron absorption.

**VITAMINS**
Fats, such as olive oil, boost the absorption of fat-soluble vitamins A, D, E, and K, and phytochemicals, all of which are found in many different foods.

**CALCIUM**
Egg yolks are a source of vitamin D, which increases the absorption of the calcium in cheese (or other calcium-rich foods) when eaten together.

**CAROTENOIDS**
Carotenoids such as lycopene, lutein, beta-carotene, and zeaxanthin are better absorbed when eaten with a source of fat, such as oil or avocado.

# WINNING COMBINATIONS

The extent to which the body can absorb a nutrient is known as bioavailability, and this varies greatly between foods. For example, around 30 per cent of the calcium in milk is absorbed but only 5 per cent of the calcium in spinach. Bioavailability is influenced by factors such as how food components interact and the nutrient's chemical form. Haem iron in meat, for instance, is more bioavailable than non-haem iron in plant-based foods. Eating certain foods together (see above) can help enhance nutrient absorption.

# GOOD PREPARATION

While some nutrients may be lost during preparation and cooking (see pp.84–85), both can increase the bioavailability of others. For example, the body absorbs more of the antioxidant lycopene from tomato purée or sauce than from fresh tomatoes as processing releases lycopene from the plant tissue. Cooking, juicing, chopping, and crushing also break down plant cell walls, increasing the amount of nutrients available for absorption. Soaking wholegrains and pulses before cooking improves the bioavailability of nutrients such as iron by reducing phytates, oxalates, and other plant chemicals – known as "antinutrients" – that bind to minerals and block their absorption.

**Chopping and crushing**
Crushed or chopped garlic and onions produce allicin, an enzyme believed to benefit heart health.

**Cooking**
By softening the tough cell walls of fruit and vegetables, cooking makes some nutrients more bioavailable.

**Juicing and processing**
As cell walls are broken down, certain nutrients in juiced or processed plant-based foods are easier to absorb.

**Soaking**
Soaking pulses and grains in water helps reduce plant chemicals that block the absorption of nutrients.

# HEALT
## FOOD
# CHOIC

# H Y

# E S

**Knowing which foods provide** the nutrients the body needs is key to choosing a healthy diet. Some foods, such as highly processed ones, offer few vitamins and minerals but contain a lot of calories, saturated fat, sugar and/or salt, while fruit, vegetables, and wholegrains supply a wide variety of the nutrients that keep us well. Modern scientific research suggests that following a balanced diet of mainly plant-based foods in the greatest possible variety – with fewer meat and dairy products (if eaten) and limited added sugar and salt – is best for optimum health.

# EAT MORE FOR LESS

Energy density is measured by the number of calories in a gram of food (kcal/g). Foods with a high fat and sugar content and little water (e.g. crisps, cake, chips, and chocolate) are energy dense and often contain few essential nutrients. Foods with a low energy density supply fewer kcal/g so can help people manage their weight. Many studies suggest that satiety (feeling satisfied after eating) is linked to the amount of food eaten, which supports the idea that large portions of low energy-dense foods, such as fruit and vegetables, typically with a high water or fibre content, may be key to achieving healthy weight loss.

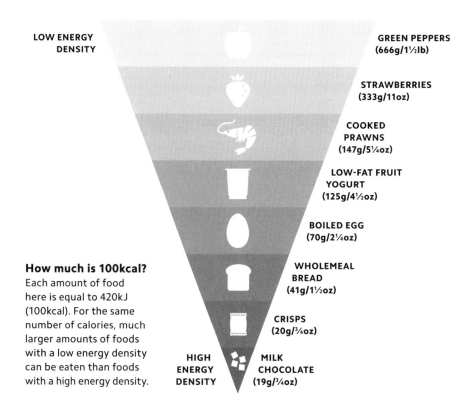

LOW ENERGY DENSITY

GREEN PEPPERS (666g/1½lb)

STRAWBERRIES (333g/11oz)

COOKED PRAWNS (147g/5¼oz)

LOW-FAT FRUIT YOGURT (125g/4½oz)

BOILED EGG (70g/2¼oz)

**How much is 100kcal?**
Each amount of food here is equal to 420kJ (100kcal). For the same number of calories, much larger amounts of foods with a low energy density can be eaten than foods with a high energy density.

WHOLEMEAL BREAD (41g/1½oz)

CRISPS (20g/¾oz)

HIGH ENERGY DENSITY

MILK CHOCOLATE (19g/¾oz)

# EVERY BITE COUNTS

Foods with plenty of nutrients relative to their calorie content
are described as nutrient dense or nutrient rich. This refers to
nutrients such as protein and healthy fats, together with fibre
and micronutrients (see p.22) that keep the many systems in the
body functioning well. Examples of nutrient-rich foods include
lean meat, chicken, fish, eggs, mycoprotein, soya, pulses, nuts,
seeds, dairy products, wholegrains, and fruit and vegetables.
The term "empty calories" is sometimes used to describe
foods that contain a lot of calories but few nutrients.

## Not like for like
These two examples provide a
similar quantity of food and a similar
number of calories. However, only
the almonds are nutrient dense.

**2 CHOCOLATE BISCUITS**
**(34g/1¼oz)**

**695kJ (166kcal)**

### Low nutrient density
High in saturated fat and
sugar and they fail to provide
significant amounts of protein,
fibre, vitamins, and minerals.

**HANDFUL OF ALMONDS**
**(30g/1oz)**

**728kJ (174kcal)**

### High nutrient density
They contain monounsaturated fat,
protein, carbohydrate, fibre, vitamin
B2, vitamin E, copper, magnesium,
manganese, phosphorus.

# A BALANCED DIET

Encouraging a healthy diet is central to public health and nutrition policies across the world. More than 100 countries offer practical recommendations adapted to suit the culture, eating habits, available foods, and nutrient needs of their populations. Known as food-based dietary guidelines, these usually offer a visual model, such as a food plate or pyramid, to illustrate food and proportion recommendations for a nutritious diet. The message is consistent. Diets should be plant-based, with plenty of fruit, vegetables, pulses, and grains. Meat and dairy foods can be eaten in moderation, but foods high in saturated fat, sugar, and salt should be limited. More countries now also include sustainability messages, enabling people to minimize the environmental impact of their food choices (see p.125).

**FRUIT AND VEGETABLES**

Eat at least five portions of a variety of fruit and vegetables each day (see pp.64–65).

**Fluids**

Drink 6–8 glasses daily of fluids such as water, low-fat milk, tea, or sugar-free drinks; limit fruit juice and smoothies to 150ml a day.

**Less nutritious foods**

Ensure that foods high in fat, sugar, and salt are only occasional treats, and eat little of them.

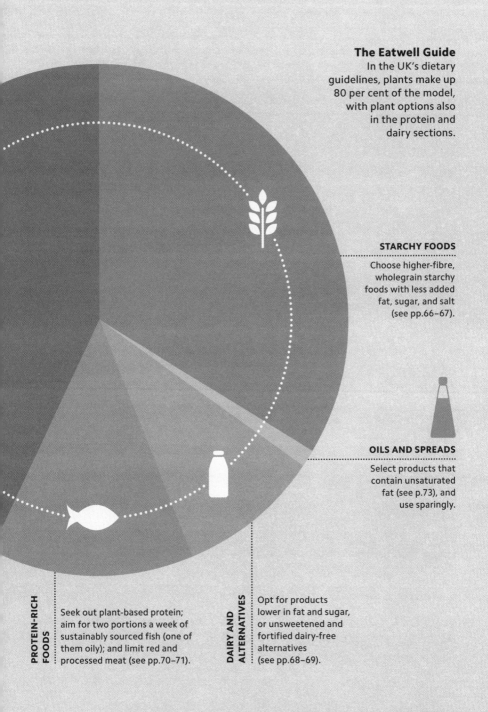

**The Eatwell Guide**
In the UK's dietary guidelines, plants make up 80 per cent of the model, with plant options also in the protein and dairy sections.

**STARCHY FOODS**
Choose higher-fibre, wholegrain starchy foods with less added fat, sugar, and salt (see pp.66–67).

**OILS AND SPREADS**
Select products that contain unsaturated fat (see p.73), and use sparingly.

**PROTEIN-RICH FOODS**
Seek out plant-based protein; aim for two portions a week of sustainably sourced fish (one of them oily); and limit red and processed meat (see pp.70–71).

**DAIRY AND ALTERNATIVES**
Opt for products lower in fat and sugar, or unsweetened and fortified dairy-free alternatives (see pp.68–69).

# PLANT POWER

Eating plenty of fruit and vegetables is the foundation of good health, as they contain vitamins, minerals, and phytonutrients, and provide dietary fibre to feed good bacteria in the gut. Numerous studies suggest it is this unique package of key nutrients (see below and opposite for examples of good sources) that maintains health and protects against heart disease, type 2 diabetes, certain cancers, and dementia, and may also help with weight management. The World Health Organization recommends at least 400g (i.e. five portions) of fruit and vegetables a day for good health, excluding potatoes, sweet potatoes, cassava, and other starchy roots. Each country translates this recommendation into their own guidelines.

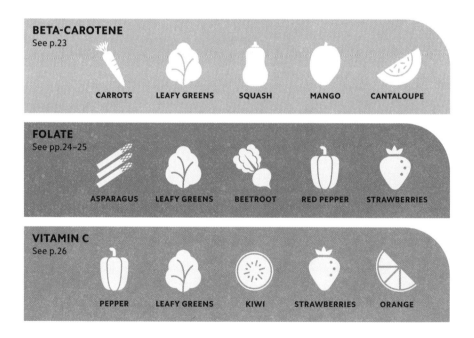

**BETA-CAROTENE**
See p.23

CARROTS　　LEAFY GREENS　　SQUASH　　MANGO　　CANTALOUPE

**FOLATE**
See pp.24–25

ASPARAGUS　　LEAFY GREENS　　BEETROOT　　RED PEPPER　　STRAWBERRIES

**VITAMIN C**
See p.26

PEPPER　　LEAFY GREENS　　KIWI　　STRAWBERRIES　　ORANGE

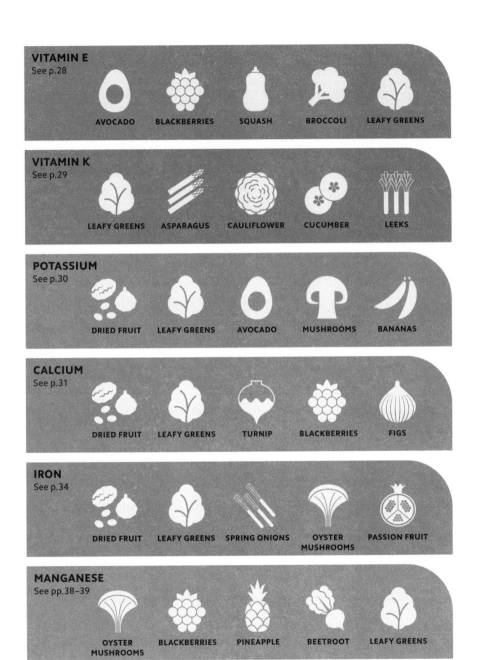

## VITAMIN E
See p.28

AVOCADO • BLACKBERRIES • SQUASH • BROCCOLI • LEAFY GREENS

## VITAMIN K
See p.29

LEAFY GREENS • ASPARAGUS • CAULIFLOWER • CUCUMBER • LEEKS

## POTASSIUM
See p.30

DRIED FRUIT • LEAFY GREENS • AVOCADO • MUSHROOMS • BANANAS

## CALCIUM
See p.31

DRIED FRUIT • LEAFY GREENS • TURNIP • BLACKBERRIES • FIGS

## IRON
See p.34

DRIED FRUIT • LEAFY GREENS • SPRING ONIONS • OYSTER MUSHROOMS • PASSION FRUIT

## MANGANESE
See pp.38–39

OYSTER MUSHROOMS • BLACKBERRIES • PINEAPPLE • BEETROOT • LEAFY GREENS

# A PRIMARY ENERGY SOURCE

Starchy foods, often called "carbs", are a primary source of nutrition and energy. Examples of starchy foods range from potatoes, plantain, and yams to grain products such as bread, rice, and pasta. The level of processing affects carbohydrate quality (see p.77). More processed or "refined" starchy foods, such as white bread, pasta, and rice, are lower in fibre and nutrients (see below). To incorporate starchy foods into a balanced, healthy diet, it is best to favour those that are high-fibre and wholegrain (e.g. wholemeal bread rather than white bread). Watching portion sizes is also advisable.

**PORTION SIZE**

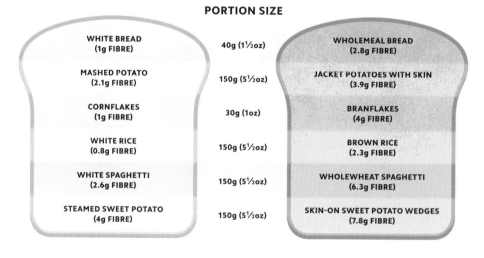

| | | |
|---|---|---|
| WHITE BREAD (1g FIBRE) | 40g (1½oz) | WHOLEMEAL BREAD (2.8g FIBRE) |
| MASHED POTATO (2.1g FIBRE) | 150g (5½oz) | JACKET POTATOES WITH SKIN (3.9g FIBRE) |
| CORNFLAKES (1g FIBRE) | 30g (1oz) | BRANFLAKES (4g FIBRE) |
| WHITE RICE (0.8g FIBRE) | 150g (5½oz) | BROWN RICE (2.3g FIBRE) |
| WHITE SPAGHETTI (2.6g FIBRE) | 150g (5½oz) | WHOLEWHEAT SPAGHETTI (6.3g FIBRE) |
| STEAMED SWEET POTATO (4g FIBRE) | 150g (5½oz) | SKIN-ON SWEET POTATO WEDGES (7.8g FIBRE) |

**Swap it out**
Pack in more nutrients, such as fibre, by choosing wholegrain or high-fibre alternatives to white starchy foods.

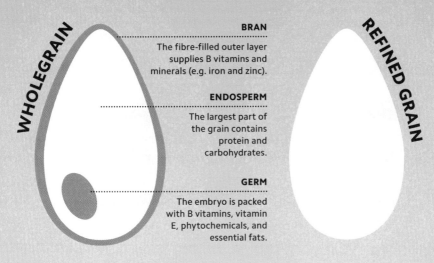

**BRAN**

The fibre-filled outer layer supplies B vitamins and minerals (e.g. iron and zinc).

**ENDOSPERM**

The largest part of the grain contains protein and carbohydrates.

**GERM**

The embryo is packed with B vitamins, vitamin E, phytochemicals, and essential fats.

**REFINED GRAIN**

**Whole lot of nutrients**

A wholegrain is made up of three parts – the bran, the endosperm, and the germ. Each part contains nutrients.

**Lost in processing**

In the refining process, the bran and the germ are removed, losing up to 75 per cent of a wholegrain's nutrients.

# UNREFINED GOODNESS

All grains start off as wholegrains that contain a range of nutrients – from fibre, vitamins, and minerals to phytochemicals, antioxidants, protein, essential fats, and starchy carbohydrates. Wholegrains are refined to create a finer texture and improve their shelf life, but this reduces their nutritional value (see above). Globally, health advice recommends replacing refined grains with wholegrains, as eating more wholegrain foods lowers the risk of heart disease, stroke, colorectal (bowel) cancer, and type 2 diabetes. While wholegrain foods usually contain more fibre, vitamins, and minerals than refined carbohydrates, they may still contain a lot of added sugar or salt, so check labels.

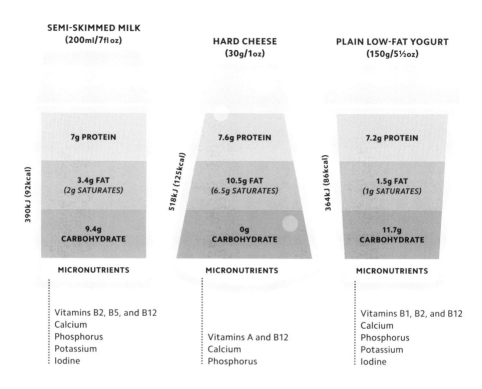

**SEMI-SKIMMED MILK**
(200ml/7fl oz)

390kJ (92kcal)

7g PROTEIN

3.4g FAT
(2g SATURATES)

9.4g
CARBOHYDRATE

MICRONUTRIENTS

Vitamins B2, B5, and B12
Calcium
Phosphorus
Potassium
Iodine

**HARD CHEESE**
(30g/1oz)

518kJ (125kcal)

7.6g PROTEIN

10.5g FAT
(6.5g SATURATES)

0g
CARBOHYDRATE

MICRONUTRIENTS

Vitamins A and B12
Calcium
Phosphorus

**PLAIN LOW-FAT YOGURT**
(150g/5½oz)

364kJ (86kcal)

7.2g PROTEIN

1.5g FAT
(1g SATURATES)

11.7g
CARBOHYDRATE

MICRONUTRIENTS

Vitamins B1, B2, and B12
Calcium
Phosphorus
Potassium
Iodine

# DAIRY NUTRITION

Milk, cheese, and yogurt are nutrient-rich and linked to good bone health, thanks to the mix of calcium, phosphorus, and protein they offer. The calcium in dairy is also more easily absorbed by the body than the calcium in most plant foods, and live yogurt contains active bacteria that supports gut health. However, full-fat dairy products are also high in saturated fat and so lower-fat dairy is often recommended to benefit heart health. The "dairy matrix" is currently a subject of scientific research, investigating how the unique physical structure and combination of nutrients and other compounds in dairy interact holistically to support health.

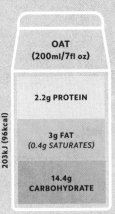

SOYA
(200ml/7fl oz)

216kJ (52kcal)

4.8g PROTEIN

3.2g FAT
(0.5g SATURATES)

1g
CARBOHYDRATE

OAT
(200ml/7fl oz)

203kJ (96kcal)

2.2g PROTEIN

3g FAT
(0.4g SATURATES)

14.4g
CARBOHYDRATE

# BEYOND THE COW

Plant-based alternatives to milk, cheese, and yogurt are suitable for vegan and dairy-free diets. They also have a lighter environmental footprint and sometimes contain fewer calories. Their nutrition depends on the plant they're made from and any added ingredients (some milk alternatives are sweetened with sugar), so it's important to check the label. Compared with dairy, most are lower in protein, calcium, iodine, and vitamins B2 and B12. However, as a result, some plant-based milk alternatives are fortified with nutrients – particularly calcium and B12 – to match cows' milk. Vitamins A and D can also be added. To meet the labelling laws in some countries, dairy-free products are called "drinks" or "alternatives", rather than "milk".

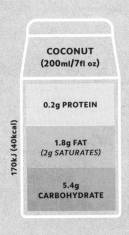

ALMOND
(200ml/7fl oz)

126kJ (30kcal)

1.1g PROTEIN

2.4g FAT
(0g SATURATES)

0.7g
CARBOHYDRATE

COCONUT
(200ml/7fl oz)

170kJ (40kcal)

0.2g PROTEIN

1.8g FAT
(2g SATURATES)

5.4g
CARBOHYDRATE

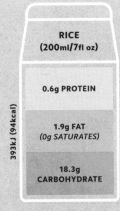

RICE
(200ml/7fl oz)

393kJ (94kcal)

0.6g PROTEIN

1.9g FAT
(0g SATURATES)

18.3g
CARBOHYDRATE

# PACKING A PUNCH

Animal proteins come from the animal itself (meat, poultry, and fish), or are produced by them (eggs and dairy products). As well as providing protein, they supply many other nutrients. For example, red meat provides iron and zinc; seafood supplies iodine and selenium; eggs and oily fish provide vitamin D. However, eating fewer animal and more plant proteins benefits both the environment and health (see p.125). The key when choosing to eat animal proteins is to do so in moderation.

| PROTEIN SOURCE | PORTION SIZE | PROTEIN CONTENT (g/PORTION) |
|---|---|---|
| BEEF | 1 STEAK (225g/8oz) | 50.6 |
| CHICKEN | 1 BREAST (150g/5½oz) | 36 |
| LAMB | 1 STEAK (150g/5½oz) | 30.3 |
| PORK | 1 LOIN CHOP (130g/4¼oz) | 28.3 |
| SALMON | 1 FILLET (130g/4¼oz) | 26.5 |
| COD | 1 MEDIUM FILLET (140g/5oz) | 24.5 |
| TURKEY | 1 THIN STEAK (100g/3½oz) | 24.4 |
| PRAWNS | 10 LARGE (100g/3½oz) | 17.6 |
| CHEESE | 1 SMALL CHUNK (30g/1oz) | 7.6 |
| PLAIN YOGURT | 1 SMALL POT (150g/5½oz) | 7.2 |
| COW'S MILK | 1 GLASS (200ml/7fl oz) | 7 |
| EGGS | 1 EGG (50g/1¾oz) | 6.3 |

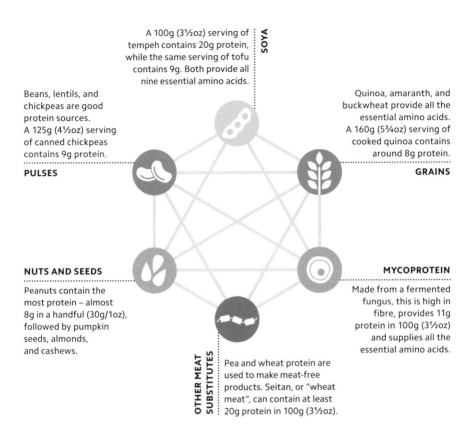

A 100g (3½oz) serving of tempeh contains 20g protein, while the same serving of tofu contains 9g. Both provide all nine essential amino acids.

Beans, lentils, and chickpeas are good protein sources. A 125g (4½oz) serving of canned chickpeas contains 9g protein.

**PULSES**

Quinoa, amaranth, and buckwheat provide all the essential amino acids. A 160g (5¾oz) serving of cooked quinoa contains around 8g protein.

**GRAINS**

**NUTS AND SEEDS**

Peanuts contain the most protein – almost 8g in a handful (30g/1oz), followed by pumpkin seeds, almonds, and cashews.

**MYCOPROTEIN**

Made from a fermented fungus, this is high in fibre, provides 11g protein in 100g (3½oz) and supplies all the essential amino acids.

**OTHER MEAT SUBSTITUTES**

Pea and wheat protein are used to make meat-free products. Seitan, or "wheat meat", can contain at least 20g protein in 100g (3½oz).

# PICKING PROTEIN

Many plants provide good amounts of protein and have the advantage over animal proteins of being low in saturated fat and often high in dietary fibre. In short, plant-based diets that are rich in variety (see above) can easily meet protein needs. In the past, the advice was to eat certain plants together, such as rice and beans, so that each meal provided a complete set of essential amino acids (see p.14). However, more recent research has shown that enjoying a range of plant foods over time can provide all the essential amino acids in adequate amounts.

# THE FUTURE OF FOOD?

The population of the world is expected to reach 9.7 billion by 2050 and the UN estimates that food production needs to double from 2018–2050 to match this rise. Therefore, alternative food sources that are nutritious, use less water and land, and incur fewer greenhouse gas emissions are being explored. Some products, designed to look and taste like meat, are made from plant proteins (e.g. soya) and fungi (e.g. mycoprotein) and are already widely available. However, alternative animal, insect, and bacterial proteins could also appear on the menu in the future.

| | | PROS | CONS |
|---|---|---|---|
|  **LAB-GROWN MEAT** | Grown from the muscle cells of an animal, this type of protein is expected to account for a quarter of all meat consumption by 2035. | Can be modified to make it healthier than regular meat. Production also costs much less than for livestock, which depends on large quantities of feed, land, and water. | As a new industry, production is still very costly. Replicating the textures of meat is difficult. Some vegans/ vegetarians may still choose to avoid eating lab-grown "meat". |
|  **SINGLE-CELL PROTEINS** | Single-cell microorganisms (e.g. algae, yeast, and fungi) are fermented, creating a nutrient-rich protein. Mycoproteins are made this way. | Can be produced rapidly, cultivated with readily available greenhouse gases, such as methane, and waste water, and grown on minimal land. | Further research and technical advances are needed so that single-cell proteins are safe to eat and can be produced on a large scale. Production costs are also high. |
|  **EDIBLE INSECTS** | Nutrient-rich insects have been part of the African, Asian, and Latin American diet for thousands of years. | Edible insects are rich in vitamins and minerals and produce more protein per unit than mammals, birds, or fish. Additionally, carbon emissions from production are very low. | Some insects may be toxic when consumed or allergenic. Sustainability is key to avoid harm to ecosystems and food chains. Also unsuitable for vegetarian and vegan diets. |

# CHOOSING FATS

All cooking fats have the same calorie value but vary in the type of fat they contain. Most oils are richer in beneficial unsaturated fats, but coconut and palm oils – like butter – have a high saturated fat content. High levels of saturated fat and industrially produced trans fats (found in partially hydrogenated vegetable oils, which some countries ban) are linked to raised cholesterol levels. In margarines and spreads, the amount and type of fat differs widely. The table below shows the approximate fat composition of 14 oils or fats.

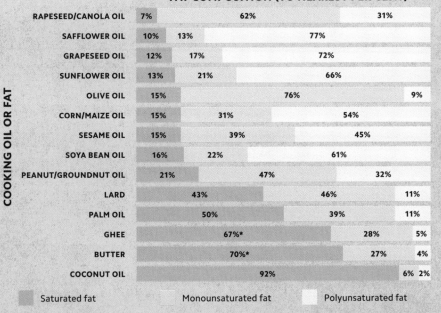

**FAT COMPOSITION (TO NEAREST PER CENT)**

COOKING OIL OR FAT

| Cooking oil or fat | Saturated fat | Monounsaturated fat | Polyunsaturated fat |
|---|---|---|---|
| RAPESEED/CANOLA OIL | 7% | 62% | 31% |
| SAFFLOWER OIL | 10% | 13% | 77% |
| GRAPESEED OIL | 12% | 17% | 72% |
| SUNFLOWER OIL | 13% | 21% | 66% |
| OLIVE OIL | 15% | 76% | 9% |
| CORN/MAIZE OIL | 15% | 31% | 54% |
| SESAME OIL | 15% | 39% | 45% |
| SOYA BEAN OIL | 16% | 22% | 61% |
| PEANUT/GROUNDNUT OIL | 21% | 47% | 32% |
| LARD | 43% | 46% | 11% |
| PALM OIL | 50% | 39% | 11% |
| GHEE | 67%* | 28% | 5% |
| BUTTER | 70%* | 27% | 4% |
| COCONUT OIL | 92% | 6% | 2% |

Saturated fat · Monounsaturated fat · Polyunsaturated fat

* Includes naturally occurring trans fats

# THE SWEET SPOT

The two main types of sugars found in food are often described as natural sugars and free sugars. Natural sugars occur naturally in fruit (and in much smaller amounts in vegetables) as fructose and in milk as lactose. Free sugars include added sugars as well as the sugars found in fruit juices, which are extracted from the cells of the fruit and so no longer come packaged with the fibre in fruit. High intake of free sugars contributes excess calories to diets, potentially causing weight gain, and increases the risk of dental decay (p.143).

**Balanced intake**
Natural sugars do not need to be limited. However, free sugars should make up no more than 10 per cent and ideally just 5 per cent of total calorie intake.

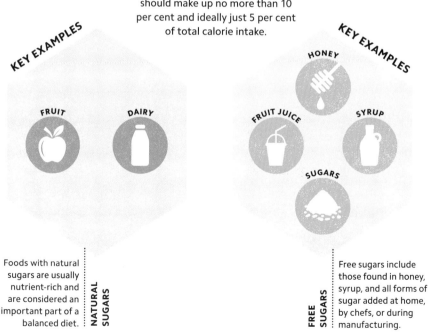

KEY EXAMPLES

FRUIT

DAIRY

KEY EXAMPLES

HONEY

FRUIT JUICE

SYRUP

SUGARS

**NATURAL SUGARS**
Foods with natural sugars are usually nutrient-rich and are considered an important part of a balanced diet.

**FREE SUGARS**
Free sugars include those found in honey, syrup, and all forms of sugar added at home, by chefs, or during manufacturing.

**A worldwide excess**
Average global salt intakes are
9–12g per day – around two
teaspoons. This is twice the
recommended upper limit of 5g.

# TOO MUCH SODIUM

In small amounts, salt – made up of sodium and chloride – is
essential to good health, including healthy nerve and muscle
function. However, average global salt intakes are far too high.
According to the World Health Organization, an upper limit
of 5g of salt per day – just one teaspoon – could prevent
2.5 million deaths a year from stroke and heart attacks. To
reduce salt intake, add less salt to food, eat fewer salty foods
(e.g. processed meats, ready meals), and use salt substitutes,
which replace some sodium with potassium.

**SWEETENERS**
The use of sweeteners as a low-calorie sugar replacement is common.

**STABILIZERS**
These additives prevent emulsified ingredients from separating.

**ANTIOXIDANTS**
Antioxidants can prevent or delay the oxidation that causes browning or makes fat rancid.

**PRESERVATIVES**
Preservatives stop food from spoiling and extend its shelf life.

**FLAVOUR ENHANCERS**
These additives bring out the flavour in food but do not add new flavours.

**EMULSIFIERS**
These additives help oil and water mix (e.g. in margarine).

**COLOURS**
Some additives change the colour of food. This is done to improve its appearance.

# IT ALL ADDS UP...

An additive is any substance (natural or artificial) added to a food item during its production – typically to keep it safe, improve its shelf life, or enhance its colour, flavour, or texture. Some common types of additive are shown above. Each type has a specific effect on certain foods. In most countries, additives must pass safety tests before their use is approved and must be listed as ingredients on food labels. Sometimes vitamins and minerals are added to food during manufacturing. This is called food fortification and is done to replace nutrients lost during processing (see opposite) or to add extra nutrients to support health.

# DEGREES OF ALTERATION

Processed foods are foods that have been altered in some way. Most foods undergo some processing to make them edible. Minimally processed foods may have been cleaned, chopped, frozen, or cooked. Highly processed foods often have added fat, sugar, and salt, along with other additives (see opposite). Though highly processed foods have been linked to health problems such as obesity, cancer, and heart disease, some experts do not believe the degree of processing is the problem. Instead, such foods may pose health risks because they are more likely to be low in micronutrients, high in energy, saturated fat, sugar, and salt, and eaten in place of more nutritious foods.

MINIMALLY PROCESSED | HIGHLY PROCESSED

JACKET POTATO — FRENCH FRIES

GRILLED CHICKEN — CHICKEN NUGGETS

APPLE — APPLE PIE

FRESH NUTS — CHOCOLATE-COATED NUTS

FRESH CARROTS — CARROT CAKE

PORK CHOP — PEPPERONI

**Processing power**
Minimally processed foods have often only been slightly altered, retaining most of their nutritional content. Highly processed foods are not inherently unhealthy but are more likely to be so.

# HEALTHY HYDRATION

Drink intake is just as important as food intake and provides around 80 per cent of the fluid in diets. Each drink has nutritional pros and cons. Water is calorie free, sugar free, and always a healthy choice. The caffeine found in many coffees, teas, and energy drinks increases alertness but can disrupt sleep. It is a mild diuretic (it increases urination) and was thought to cause dehydration, but the liquid intake from these drinks usually offsets this effect. Some high-sugar drinks (e.g. fruit juices and smoothies) contain vitamins and minerals, but others (e.g. energy and fizzy drinks) do not. Both form a significant part of daily sugar intake; teenagers in the UK may get as much as 29 per cent of their free sugars from drinks. Sugar-free drinks are far less calorific. The sweeteners they contain are safety tested, but some research suggests consuming sugar substitutes may upset the balance of gut bacteria (see pp.53–54).

| POPULAR NON-ALCOHOLIC DRINKS | PROS | CONS |
|---|---|---|
| WATER | Of all drinks, water is the top choice as it is calorie and sugar free. | Tap water is not always safe to drink in some countries. |
| MILK | Milk contains protein, calcium, phosphorus, potassium, iodine, and vitamins B2, B5, and B12. | Full-fat milk provides saturated fat. |
| COFFEE | 3–4 cups of coffee a day may lower the risk of heart disease and type 2 diabetes. | High caffeine intakes are linked to miscarriage and low birth weight in pregnancy. |

| POPULAR NON-ALCOHOLIC DRINKS | PROS | CONS |
|---|---|---|
| TEA | The antioxidants in tea may lower the risk of heart disease, stroke, and cancer. | Some teas contain tannins which inhibit absorption of non-haem iron. |
| HOT CHOCOLATE AND COCOA | Some hot chocolate and cocoa contains flavonoids (which act as antioxidants). | These drinks are often high in calories, saturated fat, and sugars. |
| FERMENTED DRINKS | Some fermented drinks (e.g. kefir and kombucha) contain probiotic bacteria (see p.55). | Fermented drinks may contain added sugar and can cause temporary bloating. |
| FRUIT JUICES AND SMOOTHIES | These drinks can count towards a person's daily intake of fruit and vegetables. | Fibre is removed with juicing and products contain free sugars (see p.74). |
| SQUASHES AND CORDIALS | Sugar-free variants of squashes and cordials add flavour to water. | Many squashes and cordials contain free sugars and little, if any, fruit juice. |
| SPORTS DRINKS | Sports drinks replenish vital minerals (electrolytes) lost in intense, prolonged exercise. | These drinks are high sugar and are not necessary for exercise of one hour or less. |
| ENERGY DRINKS AND FIZZY DRINKS | Many of these drinks come in sugar-free varieties with few or no calories. | These drinks are often high in free sugars, caffeine, and other additives. |

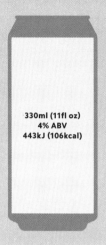

330ml (11fl oz)
4% ABV
443kJ (106kcal)

175ml (6fl oz)
13% ABV
661kJ (158kcal)

25ml (1fl oz)
40% ABV
255kJ (61kcal)

### Beer
While most beers are 6 per cent ABV or less, beer is often drunk in greater volumes than other alcoholic drinks.

### Wine
Most wines are 10–14 per cent ABV, and the stronger the wine, the more calories it will contain.

### Spirits
These drinks are commonly served in small quantities but are very strong, often ranging from 35–40 per cent ABV.

# THE RISK OF DRINKING

Alcohol is produced by fermenting grains, fruits, or vegetables. Types of alcoholic drink vary in alcohol content by volume (ABV) – the percentage of an alcoholic drink that is pure alcohol (see above). Excess alcohol consumption is linked to health problems, such as liver and heart disease, high blood pressure, and pancreatitis. It can also weaken immune function and affect the nervous system and brain. Alcohol is high in calories, and too much can also disrupt sleep and over time increase the risk of anxiety and depression. According to the World Health Organization, harmful alcohol consumption contributes to around three million deaths globally every year.

# A LONGER SHELF LIFE

Where there is warmth, moisture, oxygen, and time, bacteria, yeasts, and moulds grow. On food, these microorganisms quickly take over, in many cases rendering it inedible and harmful to health. Preservation methods (see below) can limit a food's exposure to warm, moist environments, so it can be eaten safely for longer. When products are quickly preserved after harvesting, some methods, such as freezing, lock in nutrients (such as B vitamins and vitamin C) that would be lost over time.

**Freezing or chilling**
Freezing food stops bacteria from growing, while chilling food slows bacterial growth, but neither method kills it.

**Pickling**
An acidic environment – usually vinegar – halts bacterial growth, but these acids can damage teeth.

**Canning**
Canned food is heated to kill bacteria and sealed to stop contamination. It can lose water-soluble vitamins.

**Sugaring or salting**
Although sugar and salt pose health risks, they also draw water out of food, slowing microbial growth.

**Drying or smoking**
Removing water stops bacterial growth. These methods concentrate calories and nutrients in food.

**Vacuum packing**
An airtight environment starves bacteria of oxygen. Food retains its nutrients, colour, and texture.

**Pasteurizing**
This heat treatment stops enzyme and bacterial activity, but some B vitamins are lost.

**Fermentation**
Fermented foods gain gut-friendly bacteria while food spoilage bacteria are destroyed.

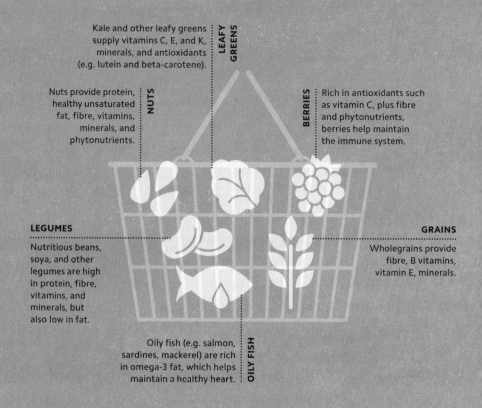

Kale and other leafy greens supply vitamins C, E, and K, minerals, and antioxidants (e.g. lutein and beta-carotene).

**LEAFY GREENS**

Nuts provide protein, healthy unsaturated fat, fibre, vitamins, minerals, and phytonutrients.

**NUTS**

**BERRIES**

Rich in antioxidants such as vitamin C, plus fibre and phytonutrients, berries help maintain the immune system.

**LEGUMES**

Nutritious beans, soya, and other legumes are high in protein, fibre, vitamins, and minerals, but also low in fat.

**GRAINS**

Wholegrains provide fibre, B vitamins, vitamin E, minerals.

Oily fish (e.g. salmon, sardines, mackerel) are rich in omega-3 fat, which helps maintain a healthy heart.

**OILY FISH**

# TRUTH OR HYPE?

The notion of health-boosting superfoods is appealing, but such foods do not really exist. There is no legal or scientific definition of a superfood. Typically, a headline proclaims (with little supporting evidence) that a single food has unrivalled disease-fighting or health-promoting properties. In such cases, identifying the food group (see above) of a hyped "superfood" can be helpful, as others in its group are likely to share most of its properties. Rather than falling for superfood hype, it is better to eat a variety of nutrient-rich foods as many are readily available, inexpensive, and tasty. The result will be a "super" diet.

# DO WE NEED SUPPLEMENTS?

Vitamin and mineral supplements can help ensure an optimal intake of nutrients, counteracting shortfalls that may result from health conditions, age, or a limited diet. They may be advised for people with increased nutrient needs or at potential risk of a nutrient deficiency; those following vegan diets, for example, should take supplemental vitamin B12.

Taking vitamin and mineral supplements cannot compensate for poor eating habits. Some studies suggest they offer little benefit unless a person has a nutrient deficiency. Unlike the nutrients in foods, which are part of a unique structure that maximizes absorption, those in supplements are isolated. High doses may inhibit nutrient absorption and even harm health.

# HEAT IT AND EAT IT

Cooking food can make it safe to eat, as the application of high temperatures to some foods (e.g. meat and fish) kills harmful bacteria (e.g. salmonella and campylobacter) that cause food poisoning. Cooking also makes foods easier to digest and more appealing in taste, texture, or appearance. As with how food is prepared (see p.57), the way it is cooked affects its nutritional content. For example, leafy greens such as broccoli or cabbage lose more of their water-soluble vitamin content (see p.22) when they are boiled than they do when steamed, fried, or roasted. For any given food, each cooking method has its advantages and disadvantages (see below and opposite).

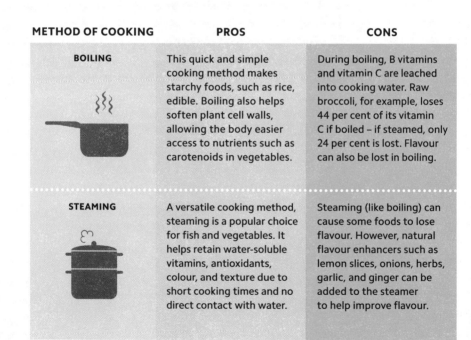

| METHOD OF COOKING | PROS | CONS |
| --- | --- | --- |
| **BOILING** | This quick and simple cooking method makes starchy foods, such as rice, edible. Boiling also helps soften plant cell walls, allowing the body easier access to nutrients such as carotenoids in vegetables. | During boiling, B vitamins and vitamin C are leached into cooking water. Raw broccoli, for example, loses 44 per cent of its vitamin C if boiled – if steamed, only 24 per cent is lost. Flavour can also be lost in boiling. |
| **STEAMING** | A versatile cooking method, steaming is a popular choice for fish and vegetables. It helps retain water-soluble vitamins, antioxidants, colour, and texture due to short cooking times and no direct contact with water. | Steaming (like boiling) can cause some foods to lose flavour. However, natural flavour enhancers such as lemon slices, onions, herbs, garlic, and ginger can be added to the steamer to help improve flavour. |

| METHOD OF COOKING | PROS | CONS |
|---|---|---|
| **FRYING**  | Frying food with fat such as oil helps the body absorb carotenoids and fat-soluble vitamins. Some oils add heart-healthy monounsaturated fats. Stir-frying and air-frying cook food quickly, which retains nutrients. | All types of frying – stir, shallow, and deep – add fat and calories. Frying above 120°C (248°F) can potentially create carcinogens: heterocyclic amines (HCAs) in meat and fish, and acrylamide in starchy foods. |
| **ROASTING**  | Cooking fish, chicken, and vegetables in parcels of foil or baking paper retains moisture and nutrients without adding fat. Using a spray oil for roasting meat, potatoes, and vegetables also reduces fat. | As with frying, roasting some foods (e.g. potatoes, meat, or fish) can form cancer-causing chemicals. To reduce these chemicals, minimize browning and reduce cooking times (as long as it is safe to do so). |
| **GRILLING AND BARBECUING**  | Fat in meat melts and drips through the grill, reducing calories and fat. Both of these cooking methods add flavour to food and are safer when food is cooked at lower temperatures and not burned. | Grilled or barbecued meat or fish form polycyclic aromatic hydrocarbons (PAHs), which are linked to cancer. Chemicals formed by frying and roasting can also be produced by grilling and barbecuing. |
| **MICROWAVING**  | Little water and fast cooking mean nutrients are retained (e.g. peas microwaved from frozen contain 30 per cent more vitamin C than if boiled). Microwaving is also useful for reheating leftovers. | Some non-microwave-safe plastic containers can transfer harmful chemicals into food when microwaved. If food is microwaved with too much water or not covered, it is easy to overcook or recreate boiling. |

# AGES
AND
# STAG

E  S

**From infancy to old age,** it is important to have a healthy, varied diet that adapts to the body's evolving nutritional needs. During childhood and teenage years, energy and nutrient requirements are high to support rapid growth and development. Pregnancy and lactation also increase the need for certain nutrients, as do changes that take place in our middle and later years. Knowing what the body requires at different stages helps maintain wellness throughout life.

## INFANTS

The BMR of babies
and children is
1.5–2 times higher
than adults.

## ADULTS

Being larger and having
a higher proportion of
muscle, men tend to have
a higher BMR than women.

BMR increases because of
increased body mass and
physiological changes
associated with
pregnancy. BMR also
increases during
breastfeeding.

## PREGNANCY AND
## BREASTFEEDING

Muscle uses more
energy to function
than fat. Muscle mass
often reduces later
in life, leading to a
reduction in BMR.

## OLDER ADULTS

# A BURNING ISSUE

The term metabolism describes the calorie-burning chemical
reactions that occur within the body's cells to maintain life.
The speed of these processes, the basal metabolic rate (BMR),
is influenced by factors such as genetics, body size and
weight, and age (see pp.12–13). BMR is highest at birth,
as babies and young children have increased energy
requirements for growth and development. Muscle is
metabolically more active than fat so burns more calories.
This means the loss of muscle mass associated with aging
contributes to a fall in BMR. The most effective way to
boost metabolism is to build muscle through exercise.

# EARLY YEARS NUTRITION

From birth to 12 months, babies grow faster than at any other stage in life. Milk (breast or formula) should be the only source of nutrition for the first six months. After this time, a baby's stores of some nutrients (e.g. iron) start to run out, and foods must be introduced to replenish them. This process was once referred to as weaning, but many health professionals now call it complementary feeding. Introducing solid foods helps develop the muscles necessary for chewing and speech. During this time, it is important to gradually increase the amount, variety, and texture of foods in a baby's diet.

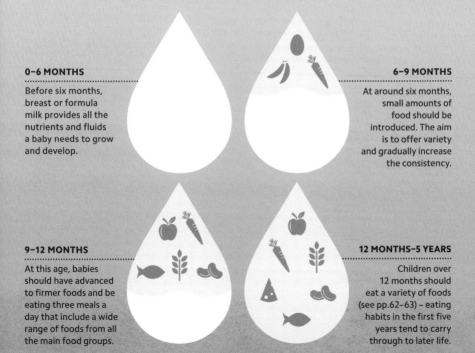

**0–6 MONTHS**

Before six months, breast or formula milk provides all the nutrients and fluids a baby needs to grow and develop.

**6–9 MONTHS**

At around six months, small amounts of food should be introduced. The aim is to offer variety and gradually increase the consistency.

**9–12 MONTHS**

At this age, babies should have advanced to firmer foods and be eating three meals a day that include a wide range of foods from all the main food groups.

**12 MONTHS–5 YEARS**

Children over 12 months should eat a variety of foods (see pp.62–63) – eating habits in the first five years tend to carry through to later life.

# CREATING GOOD HABITS

It is important to form healthy eating habits (see below) at an early age in order to fuel development, build strong bones, and reduce the risk of health problems in later life. This period sees slow, steady growth but energy, protein, and micronutrient requirements are relatively high compared to body size and increase as children get older. Protein is especially important for the growth, maintenance, and repair of body tissues. Snacking habits often develop during this stage of life, so it is crucial that the snacks provided are healthy and nutrient-rich (see p.116).

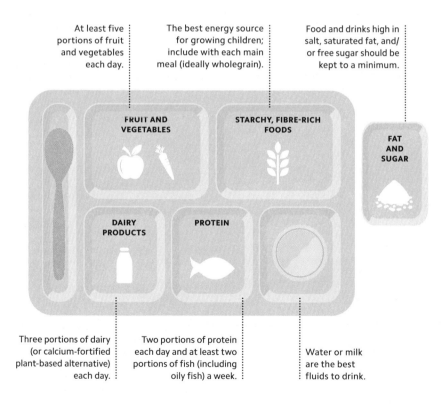

At least five portions of fruit and vegetables each day.

The best energy source for growing children; include with each main meal (ideally wholegrain).

Food and drinks high in salt, saturated fat, and/or free sugar should be kept to a minimum.

**FRUIT AND VEGETABLES**

**STARCHY, FIBRE-RICH FOODS**

**FAT AND SUGAR**

**DAIRY PRODUCTS**

**PROTEIN**

Three portions of dairy (or calcium-fortified plant-based alternative) each day.

Two portions of protein each day and at least two portions of fish (including oily fish) a week.

Water or milk are the best fluids to drink.

# GROWING UP

## KEY NUTRIENTS FOR TEENAGERS

### Iron
Crucial during all periods of rapid growth (see p.34 for sources), but menstruation increases iron requirements even further.

### Iodine
Low intakes could delay growth and affect cognitive function (see p.37). Teens who do not consume fish or milk – the main sources of iodine in the diet – should consider supplements.

### Omega-3 fats
Research suggests that omega-3s play a key role in brain function. Good sources include oily fish, walnuts, and omega-3-enriched foods.

### Calcium
For bone health (see p.108), dairy and/or fortified non-dairy sources (see p.31) are essential. Teens who are vegan may require a supplement.

The growth spurt and developmental changes that accompany the teenage years increase energy and nutrient requirements. For some nutrients (e.g. calcium, phosphorus, and iron), requirements are higher for teenagers than adults. However, growing independence can mean that parents or carers have less control over what teens eat. Pressure to conform to a certain body image or an interest in animal welfare or environmental issues may also influence food choices at this age. Teens often have very low intakes of many key nutrients (see above) and can be at risk of deficiencies.

# EATING AND LIVING WELL

A healthy, balanced diet (see pp.62–63) is important for all ages and groups, and adults are no exception. For many, however, it is a time when work and family life take centre stage, which means that healthy eating often gets forgotten. A healthy diet will help the body cope better with a busy life and stress and reduce the risk of heart disease, obesity, type 2 diabetes, high blood pressure, and digestive disorders. Surveys show that many adults fail to meet the recommended targets for fruit, vegetables, dietary fibre, oily fish, and some key vitamins and minerals (vitamins A and D, and folate, and minerals calcium and iron) and at the same time consume too much saturated fat, sugar, and salt.

**AVOID**
Drinking alcohol can adversely affect ovulation and sperm motility.

**LIMIT**
Some studies suggest high intakes of caffeine can negatively impact female fertility.

**LIMIT**
A diet of highly processed foods and foods high in saturated fats and trans fats is linked to infertility.

**EAT MORE**
Foods high in monounsaturated fat can boost female fertility, while omega-3 fats may improve sperm quality.

**EAT MORE**
Vegetables, wholegrains, and pulses contain many essential micronutrients and antioxidants that help boost fertility.

# EATING TO CONCEIVE

Couples who adopt a healthy diet and lifestyle for at least three months before trying to conceive maximize their chances of success. Deficiencies of nutrients (e.g. vitamin D, iron, zinc, and selenium) can reduce fertility so eating a healthy diet which contains a variety of fruit and vegetables, wholegrain cereals, plant proteins, and healthy fats is important. A folic acid supplement is also recommended for people who are trying to become pregnant as it can reduce the risk of birth defects. Achieving a healthy weight is also important (see p.134). A certain level of body fat is needed for ovulation so being under or overweight can negatively impact fertility.

# NUTRITION FOR TWO

## AVOID

RAW OR UNDERCOOKED MEAT, POULTRY, AND SHELLFISH

LIVER, PÂTÉ, AND MULTIVITAMIN SUPPLEMENTS CONTAINING VITAMIN A

UNPASTEURIZED DAIRY PRODUCTS

RAW OR LIGHTLY COOKED EGGS

FISH HIGH IN MERCURY (E.G. SHARK, SWORDFISH, MARLIN)

ALCOHOL

## LIMIT

CAFFEINE
(NO MORE THAN 200MG A DAY;
E.G. ONE MUG OF FILTER COFFEE
OR THREE CUPS OF TEA)

OILY FISH
(NO MORE THAN
TWO PORTIONS A WEEK)

## EAT/DRINK

FRUIT AND VEGETABLES, PARTICULARLY LEAFY GREENS

IRON-RICH FOODS (E.G. LEAN RED MEAT, FORTIFIED CEREALS, PULSES)

FIBRE-RICH FOODS (E.G. BEANS, WHOLEGRAINS, VEGETABLES, FRUIT)

PASTEURIZED DAIRY FOODS

PLENTY OF WATER

A healthy diet during pregnancy will give a baby the best start in life. Requirements for nutrients (e.g. calcium and iron) increase during pregnancy, but the body also adapts during this time to absorb more of these from food. Meanwhile, calorie needs increase only slightly. This means that diet during pregnancy needs to contain nutrient-dense foods (see p.61). Recommendations vary between countries, but the advice in the UK is that people who are pregnant and a healthy weight require an additional 840kJ (200kcal) a day in the final trimester. Ideally before conception and for the first 12 weeks of pregnancy, a daily 400mcg folic acid supplement should be taken, along with a daily 10mcg vitamin D supplement for the entire pregnancy. Those following vegetarian and vegan diets may also need to take a supplement of iodine, omega-3 fats, and vitamin B12.

**BREAST BENEFITS**

Breastfeeding, for those who can, brings benefits for both mother and baby. It can help to protect babies against some short- and long-term illnesses and diseases (e.g. infections and allergies, as well as heart disease, type 2 diabetes, and obesity later in life). For women, breastfeeding also reduces the risk of breast and ovarian cancer, obesity, and cardiovascular disease. Breastfeeding increases requirements for fluids, energy, protein, and most vitamins, as well as calcium, phosphorus, magnesium, zinc, copper, and selenium. Therefore, it is essential that breastfeeding goes hand in hand with a nutrient-rich diet (see pp.62–63); especially wholegrains, dairy or calcium-fortified dairy alternatives, and a variety of proteins, including oil-rich fish.

**REDUCE ALCOHOL AND CAFFEINE**
Alcohol and caffeine can exacerbate symptoms (e.g. anxiety, hot flushes, sleep problems). Alcohol can also contribute to weight gain.

**PROTECT THE HEART**
Oestrogen encourages the body to produce HDL "good" cholesterol that protects the heart (see p.104). After the menopause, this protection is lost.

**EAT REGULARLY**
Dips in blood sugar levels can intensify menopausal symptoms, such as irritability, tiredness, poor concentration, and headaches.

**EAT SOYA**
Soya beans and related products contain phytoestrogens, which have a mild oestrogen-like effect that may help reduce hot flushes and night sweats.

**MANAGE WEIGHT**
Metabolism slows only marginally at this time but, when combined with symptoms that affect exercise and eating habits, weight gain is common.

**PROTECT BONE HEALTH**
Oestrogen decline accelerates the loss of calcium from bones. Include calcium and other bone-friendly nutrients (see p.108).

# MANAGING THE CHANGE

Informed dietary choices can help alleviate some of the symptoms that often accompany the perimenopause (the period leading up to the menopause) and the menopause itself. Although not all women experience symptoms, the decline in oestrogen that occurs at this time (usually between the ages of 45 and 55) increases the risk of other health issues, such as heart disease, osteoporosis, and weight gain. However, making lifestyle changes, such as regular exercise, and following healthy eating advice (see above) can help reduce the impact of these problems.

# A RIPE OLD AGE

The physiological and psychological changes that occur with aging can have a big impact on nutritional status and requirements. Many people find their appetite decreases in later life. However, the need for nutrients remains the same or can increase, so it becomes even more important to have a nutrient-dense diet (see p.60). A low protein intake can exacerbate muscle loss (sarcopenia), which increases the risk of falls. Calcium and vitamin D are especially important for protecting bones; B vitamins for cognitive health; omega-3 fats for a healthy heart; and fibre to prevent the digestive system from getting sluggish.

**STAY HYDRATED**
Drinking prevents dehydration, which is a risk as the body becomes less sensitive to thirst signals with age.

**MANAGE WEIGHT GAIN**
A diet of nutrient-dense rather than energy-dense foods prevents excessive weight gain, which often accompanies the natural loss of muscle mass and drop in metabolic rate that occurs with age.

**MAKE MEALS COUNT**
Basing meals, snacks, and drinks around nutrient-dense foods can help ensure the diet provides adequate nutrition, even when appetite is reduced.

> With age, the body becomes less efficient at processing some micronutrients.

**PREVENT MUSCLE LOSS**
A dramatic loss of muscle mass (sarcopenia) increases frailty, which further limits mobility and is often accompanied by rapid weight loss.

# TOP -
# TO - TOE
HEALTH

**Diet affects every part of the body,** from eyes, teeth, and skin to heart, bones, and joints. At the centre is the gut microbiome, which not only affects the health of the digestive system, but also interacts with other parts of the body, such as the immune system and brain. Such interplay reveals how nutrients that support one part of the body can also influence another. It underscores the importance of a nutrient-packed diet for whole-body health.

# FEEDING
# THE BRAIN

The brain depends on good nutrition (see below) and uses glucose as its preferred fuel for energy-hungry brain cells. The omega-3 fatty acid DHA is found in cell membranes and is essential for transmitting information between nerve cells (neurons) in the brain and spinal cord. Iodine, iron, and zinc support normal cognitive function (i.e. all the processes that affect memory, attention, thinking, reasoning, and problem-solving), and staying hydrated is key as the brain is around 75 per cent water. Choline, a nutrient in eggs, dairy, meat, chicken, fish, beans, nuts, and leafy greens is also important as it makes acetylcholine, a neurotransmitter that aids brain and nerve function, and also memory.

## KEY NUTRITION

**DHA**
Important for maintaining normal brain function. Found in oily fish.

**Iodine**
Essential for cognitive function throughout life. See p.37 for sources.

**Iron**
Contributes to normal cognitive function. See p.34 for sources.

**B vitamins**
Support cognitive function. See pp.24–25 for sources.

**KEY NUTRITION**

**Protein**
Helps the body
produce keratin,
which strengthens
hair. See pp.70–71
for sources.

**Biotin (vitamin B7)**
May play a role in
healthy hair growth.
See pp.24–25
for sources.

**Selenium**
Keeps hair healthy,
but too much (from
supplements) may
cause hair loss.
See p.36 for sources.

**Zinc**
Supports hair health,
especially hair
growth. See p.35
for sources.

# NOURISHING THE ROOTS

Hair condition directly reflects diet quality as every hair follicle
needs a supply of energy and nutrients (see above) to feed the
cells of the hair root growing within it. When nutrient intakes
are inadequate – perhaps as a result of illness or extreme dieting –
hair growth is disrupted and, after around three months, thinning
or hair loss may occur. Eating protein-rich foods helps ensure the
body's production of the protein keratin, which gives hair its
structure and strength; insufficient keratin leaves hair brittle and
weak. Iron-deficiency anaemia (see p.152) can also cause hair
loss, so it is advisable to eat plenty of iron-rich foods.

# LOOKING AFTER EYES

A vision-friendly diet, rich in key nutrients (see below),
can protect the eyes and reduce the risk of cataracts
and age-related macular degeneration (AMD), both of which
can impair sight. Vitamin A is essential for vision in dim light.
The antioxidant power of lutein, zeaxanthin, and vitamins C
and E may prevent the oxidation that causes the lens to become
cloudy and results in cataracts. Zinc is also important as it
is highly concentrated in the eye, and works with vitamin A
to support good vision. Maintaining healthy blood glucose
(see p.135) and cholesterol (see p.136) levels also helps prevent
damage to the blood vessels that nourish the eyes.

## KEY NUTRITION

**Vitamin A**
Transforms into
a pigment in the
retina that aids
vision in dim
light. See p.23
for sources.

**Lutein and
zeaxanthin**
Found in all leafy
greens, these
potent antioxidants
seem to protect
against AMD.

**Vitamin C**
May protect
against and slow
the progression
of cataracts.
See p.26
for sources.

**DHA**
An omega-3 fat
needed for eye
development in babies
and good vision
throughout life. Oily
fish is the best source.

# MAINTAINING THE MOUTH

A healthy mouth is about ensuring teeth and gums remain strong and this can be achieved with good nutrition (see below). Calcium, phosphorus, and magnesium maintain structure and strength in the jawbone and teeth, while fluoride is important for strengthening tooth enamel. Vitamin C keeps the connective tissues healthy, and vitamins A, B2, B3, and B7 maintain the moist lining of the mouth (the mucous membrane). Mouth ulcers can occur when immunity is compromised, for example when stressed or ill, or as a result of an iron or vitamin B12 deficiency. Frequently exposing teeth to sugar is one of the main causes of tooth decay (see p.143).

## KEY NUTRITION

**Calcium and phosphorus**
Keep the jawbone strong and give teeth their structure and strength. See pp.31–32 for sources.

**Magnesium**
Hardens the enamel on the tooth surface, which helps to protect against decay. See p.33 for sources.

**Vitamin C**
Keeps the connective tissues strong and protects against gum disease. See p.26 for sources.

**Vitamin D**
Aids the absorption of calcium and phosphorus from food. See p.27 for sources.

# PROTECTING THE HEART

Good nutrition is essential for heart health as key nutrients (see below) can help keep the heart and circulation working optimally and protect against disease. Nitrate-rich foods (e.g. beetroot and leafy greens) produce a gas called nitric oxide in the blood and this relaxes blood vessels, improving oxygen and nutrient flow around the body and helping to lower blood pressure. Polyphenols (found in many plant foods) have a similar effect and also provide antioxidant support. Saturated fats and trans fats are both associated with "bad" LDL cholesterol (see p.136) and consequent cardiovascular disorders.

## KEY NUTRITION

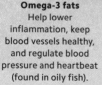

**Omega-3 fats**
Help lower inflammation, keep blood vessels healthy, and regulate blood pressure and heartbeat (found in oily fish).

**Magnesium**
Needed for muscle contractions that give a steady heartbeat. See p.33 for sources.

**Vitamin C**
Helps to make collagen, which is vital for healthy blood vessels. See p.26 for sources.

**Potassium**
Works with sodium to control blood pressure. See p.30 for sources.

## KEY NUTRITION

**Iron**
Part of haemoglobin, which transports oxygen around the body. See p.34 for sources.

**B vitamins**
Vitamins B2, B6, B9, and B12 are vital for making healthy red blood cells. See pp.24–25 for sources.

**Water**
As blood is 90 per cent water, staying hydrated prevents a drop in blood volume.

**Vitamin C**
Helps the body absorb iron from plant foods. See p.26 for sources.

CIRCULATING NUTRIENTS

Blood carries nutrients to cells but also requires vitamins and minerals itself in order to function effectively. Without key nutrition (see above), blood lacks red blood cells, which can cause anaemia (see p.152) and, if under-hydrated, becomes thick, making it harder for the heart to pump it around the body. Calcium and vitamin K are also essential nutrients for helping blood to clot, while copper helps to transport iron around the body, so a copper deficiency can also lead to anaemia.

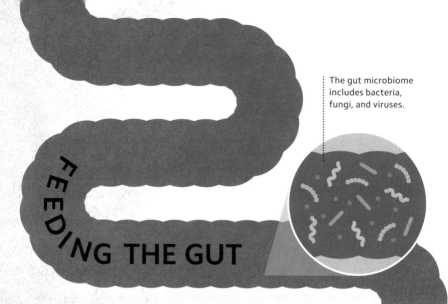

The gut microbiome includes bacteria, fungi, and viruses.

# FEEDING THE GUT

The large intestines – often referred to as the gut – are home to about 100 trillion microbes (known as the gut microbiome, see pp.54–55). Research suggests that the balance of gut microbes affects many aspects of health. A varied, fibre-rich diet provides digestive benefits (see pp.20–21) and helps maintain a healthy balance between "good" and potentially harmful gut bacteria. When fibres are fermented by gut bacteria they produce short-chain fatty acids (SCFAs), which feed and strengthen the gut lining. Prebiotic fibres feed "good" bacteria, while probiotics boost gut diversity.

## KEY NUTRITION

**Prebiotics**
Indigestible fibres found in many plant foods feed "good" gut bacteria.

**Probiotics**
Fermented products (e.g. kimchi) add beneficial bacteria to the gut microbiome.

**Polyphenols**
Gut bacteria transform these compounds, found in all plant-based foods, into health-boosting substances.

**Water**
Water keeps food moving through the intestines and helps to prevent constipation.

## KEY NUTRITION

**Vitamin C**
Protects the immune
system by neutralizing
cell-damaging free
radicals. See p.26
for sources.

**Vitamin D**
Reduces
inflammatory
responses and helps
regulate immune
system function.
See p.27 for sources.

**Selenium**
Helps strengthen the
body's response to
infection and makes
new immune cells.
See p.36 for sources.

**Zinc**
Produces new
immune cells and aids
communication
between them.
See p.35 for sources.

The immune system is
a complex network of cells, tissues, and
organs which destroys toxins, protects against germs,
and fights cell changes that cause illnesses, such as
cancer. Eating a varied, plant-rich diet contributes to a
healthy immune system by supplying key nutrients
(see above) and additional vitamins (A, B6, B9, B12)
and minerals (iron and copper). These nutrients
work in many ways, such as helping to produce
immune cells and supporting communication
between them, and boosting response to
infection. Studies also suggest that the
key to a strong immune system is
a healthy gut microbiome
(see opposite).

BOOSTING IMMUNITY

# BUILDING BONES

Strong, healthy bones rely on good nutrition (see below) and the most important nutrient here is calcium. The first 30 years of life are key for bone building, particularly during the teenage growth spurt, so the more calcium deposited in the bones at this time, the more protected they are against osteoporosis in later life (see p.150). Vitamin D helps the body absorb and use calcium, while vitamin C is needed to make collagen, a protein that is essential for bone formation. However, very high intakes of retinol (vitamin A) – mainly from supplements or frequently eating liver or liver products – cause bones to lose strength by reducing mineral density.

## KEY NUTRITION

**Calcium**
Essential for healthy bone growth. See p.31 for sources.

**Vitamin D**
Helps the body absorb calcium, promoting healthy bone growth. See p.27 for sources.

**Vitamin K**
Helps to build and heal bones by activating a protein hormone called osteocalcin. See p.29 for sources.

**Magnesium**
Helps to harden and strengthen bones. See p.33 for sources.

**Vitamin C**
Supports healthy
connective tissue.
See p.26 for sources.

**Manganese
and copper**
Support normal
formation of
connective tissue.
See pp.38–39
for sources.

# SUPPORTING JOINTS

Joints are the areas in the
body where two bones meet, to allow
movement. Each joint is supported by tendons
(connecting muscle to bone), ligaments (joining
bones to other bones), and cartilage (padding
the ends of bones and cushioning between joints).
Many nutrients (see right) support these body
parts. Vitamin C, manganese, and copper
are needed to form connective tissue,
which is a key component of each.
Furthermore, as joints are part
of the musculoskeletal system,
diets must also promote
healthy bones and
muscles (see pp.108
and 110).

**Vitamin K**
Works with protein
in cartilage to
support its function.
See p.29 for sources.

**Omega-3 fats**
May ease
inflammation. Oily
fish is the best source.

# POWERING MUSCLES

Almost half of all protein in the body is contained in muscles, so a protein-rich diet is essential for muscle growth and repair. Calcium, magnesium, and potassium enable muscles to contract and relax, while vitamin D helps them to function at their best. If the latter is lacking, muscle weakness and pain can occur. In order to build muscle, adequate protein intake needs to be combined with resistance exercise (e.g. lifting weights) – and sufficient calories and carbohydrates after exercise to refuel glycogen stores so that muscle tissue can regenerate.

## KEY NUTRITION

**Protein**
Essential for building and maintaining muscle mass. See pp.70–71 for sources.

**Calcium**
Helps to transmit nerve impulses, which trigger muscle contraction. See p.31 for sources.

**Magnesium**
Essential for muscle relaxation and preventing cramp. See p.33 for sources.

**Potassium**
Needed for muscle contraction and overall function. See p.30 for sources.

> When the skin is exposed to sunlight, the body can create vitamin D.

# SUSTAINING THE SKIN

Certain nutrients can provide the foundation for healthy skin, so it is able to fulfil its role as a protective barrier against viruses and bacteria. Key nutrients (see below) not only regenerate and repair damaged skin but also have antioxidant powers, eliminating cell-damaging free radicals that are linked to premature aging. Good fats also support healthy skin – omega-3 fats, for example, reduce inflammation, while monounsaturated fats may protect against skin aging. Limiting sugar may also prevent the weakening of elastin and collagen – essential proteins that give shape, structure, and firmness to skin.

## KEY NUTRITION

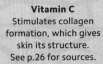

**Vitamin A**
Assists with the growth and repair of skin tissue. See p.23 for sources.

**Vitamin C**
Stimulates collagen formation, which gives skin its structure. See p.26 for sources.

**Vitamin E**
Aids wound healing and prevents scarring. See p.28 for sources.

**Zinc**
Helps ease inflammation and supports wound healing. See p.35 for sources.

# EATING
## AND DIE

# HABITS

## TS

**What we eat,** when, how often, how much, and how fast all contribute to our eating habits. These are shaped by many factors, including family and friends, individual preferences and beliefs, and financial and cultural influences. Eating habits are established early in life but develop with age and changing circumstances. Our diet is also shaped by geography (which dictates the availability of foods), our overall health, any medical conditions, and environmental concerns. Together, these factors significantly influence the food we choose to eat.

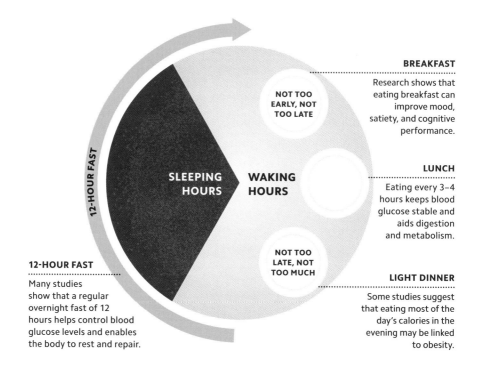

**BREAKFAST**

Research shows that eating breakfast can improve mood, satiety, and cognitive performance.

**LUNCH**

Eating every 3–4 hours keeps blood glucose stable and aids digestion and metabolism.

**NOT TOO EARLY, NOT TOO LATE**

**SLEEPING HOURS**

**WAKING HOURS**

**NOT TOO LATE, NOT TOO MUCH**

**12-HOUR FAST**

**12-HOUR FAST**

Many studies show that a regular overnight fast of 12 hours helps control blood glucose levels and enables the body to rest and repair.

**LIGHT DINNER**

Some studies suggest that eating most of the day's calories in the evening may be linked to obesity.

# TIME TO EAT

Eating patterns can vary and depend on several factors, including culture, habit, hunger, stage of life, and work patterns. However, for most people, the majority of calories are consumed in a 12-hour period between waking and bed time. Some studies suggest that nutrient intakes can be related to eating patterns. The most common finding is that skipping breakfast results in a lower intake of key nutrients and poorer overall diet quality. The impact of snacking (p.116) and frequency of eating is unclear. For some groups, like young children, older adults, or anyone with a small appetite, eating little and often can be a beneficial pattern of eating. However, snack foods are often high in fat, sugar, and/ or salt and do not provide the same nutritional value as a meal.

# A NEGATIVE CYCLE

Skipping meals can have an adverse effect on both long-term health and day-to-day mental health and physical performance. In one large US study of adults over 40, eating only one meal a day was linked to higher rates of heart disease and mortality. Another study showed that young people who skipped breakfast were more likely to experience stress or depression than those who ate breakfast. Glucose is the number one fuel for the brain so, when blood glucose levels drop, it often leads to cravings for fatty and sugary snacks that promise instant energy, and a pattern of making poor food choices can quickly become established.

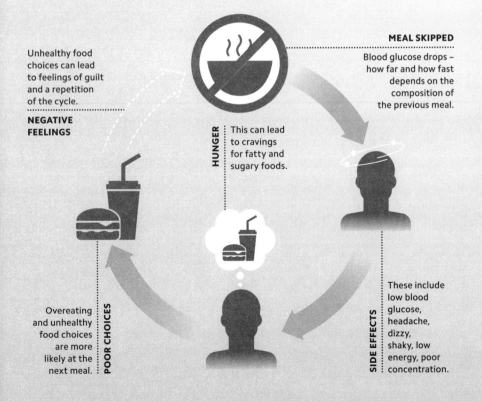

**MEAL SKIPPED**
Blood glucose drops – how far and how fast depends on the composition of the previous meal.

Unhealthy food choices can lead to feelings of guilt and a repetition of the cycle.
**NEGATIVE FEELINGS**

**HUNGER**
This can lead to cravings for fatty and sugary foods.

**SIDE EFFECTS**
These include low blood glucose, headache, dizzy, shaky, low energy, poor concentration.

**POOR CHOICES**
Overeating and unhealthy food choices are more likely at the next meal.

# A QUICK BITE

Defined as any food eaten between meals, a snack can make a positive contribution to a day's nutrient intake. It can prevent a dip in blood glucose levels, provide a useful energy boost, and help curb hunger, reducing the risk of overeating at the next meal. For those with a small appetite, snacks can help maintain good nutrition. But many snack foods are processed and high in calories, saturated fat, sugar, and salt, so eating too many can contribute to health problems such as obesity and tooth decay. Healthier snacks (see below) are lower in these ingredients, contain fewer calories, and supply key nutrients, such as protein, fibre, vitamins, and minerals.

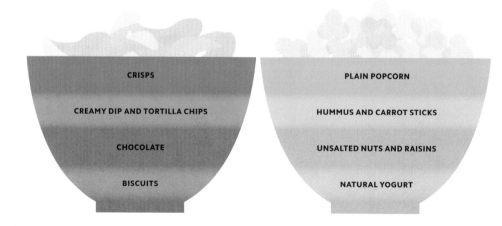

CRISPS

CREAMY DIP AND TORTILLA CHIPS

CHOCOLATE

BISCUITS

PLAIN POPCORN

HUMMUS AND CARROT STICKS

UNSALTED NUTS AND RAISINS

NATURAL YOGURT

**Healthier choices**
Stocking up on foods in the green bowl, rather than those in the red bowl will ensure there are always healthier snacks to hand.

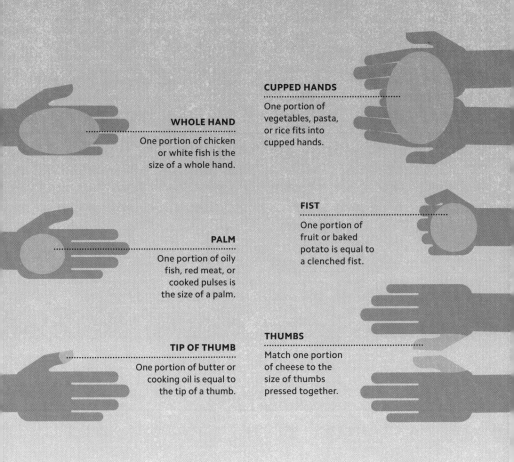

**CUPPED HANDS**

One portion of
vegetables, pasta,
or rice fits into
cupped hands.

**WHOLE HAND**

One portion of chicken
or white fish is the
size of a whole hand.

**FIST**

One portion of
fruit or baked
potato is equal to
a clenched fist.

**PALM**

One portion of oily
fish, red meat, or
cooked pulses is
the size of a palm.

**THUMBS**

Match one portion
of cheese to the
size of thumbs
pressed together.

**TIP OF THUMB**

One portion of butter or
cooking oil is equal to
the tip of a thumb.

# HOW MUCH?

Portion control is key to a balanced diet and weight management
(see p.134). Appropriate portion sizes vary based on factors such as
age, gender, and activity level. For a simple guide, hands provide a
good starting point for many foods (see above). This approach works
because people with smaller bodies, who often require smaller
portions, tend to have smaller hands. Use appetite as a guide, but
aim for a daily intake of five portions of fruit and vegetables; 3–4
portions of starchy, fibre-rich foods; 2–3 portions of dairy or
alternatives (see pp.68–69); and two portions of protein-rich foods.

# EATING TRIGGERS

People eat for many reasons besides hunger. Eating often has more to do with habit or with satisfying an emotional need than supplying the body with the energy and nutrients it requires. This can lead to an energy surplus in a diet and result in weight gain. People often use food as a reward, to celebrate, or to combat emotions such as boredom, loneliness, anxiety, or sadness. Certain people, places, and situations can also act as triggers for non-hunger eating. Keeping a food and mood diary can help to identify these triggers (see below), so that strategies can be adopted to recognize and minimize non-hunger eating.

## IDENTIFYING THE TRIGGER

### Emotions
Any emotion can trigger the urge to eat. To avoid emotional eating, choose to do something else until the urge passes, such as phoning a friend or listening to music.

### Places
In some places, snacking is common (e.g. in the workplace). Swapping high-sugar treats for healthier alternatives (e.g. fruit) is a helpful solution.

### People
It can be hard to eat healthy or stick to appropriate portion sizes when food is prepared by others. To combat this, advise people of your boundaries.

### Social events
Food and social events often go hand in hand – especially with alcohol (see p.80). Eat a healthy snack ahead of time to avoid arriving hungry.

# PAUSING FOR THOUGHT

## LISTEN TO YOUR BODY

Practise recognizing when you feel full.

## EAT SLOWLY AND CHEW PROPERLY

It takes 20 minutes for the stomach to send signals to the brain that it is full.

## FOCUS ON FOOD

Sit at a table and give food your full attention. Savour every mouthful.

**DO NOT MULTI-TASK**

Avoid distractions like the TV, computer, or phone.

Mindful eating encourages a person to be in the moment and sensually aware of the food and drink they consume. Studies show that mindful eating can aid digestion, help regulate appetite, make meals more satisfying, and promote a healthier relationship with food. Although mindful eating is not focused on weight loss, the two often go hand in hand, as being aware of what we are consuming and eating at a slower pace allows the body to more easily recognize when it is full.

**ENGAGE THE SENSES**

Notice the colour, flavour, texture, and smell of the food you eat.

# FOOD LABELS

Labels on prepacked food can provide important information about a food's ingredients, energy value (kJ/kcal), and nutrients – sometimes including the vitamins and minerals it contains. Labels may also indicate the amount of key nutrients in a typical portion of the food – usually expressed as a percentage of the guideline or recommended daily amount. By law, a label must also highlight major allergens in a food, such as dairy, nuts, or wheat – enabling people with allergies to determine which foods are safe for them to eat.

| | LOW | HIGH |
|---|---|---|
| TOTAL FAT | 3g or less/100g | More than 17.5g/100g |
| SATURATES | 1.5g or less/100g | More than 5g/100g |
| TOTAL SUGAR | 5g or less/100g | More than 22.5g/100g |
| SALT (SODIUM) | 0.3g or less/100g (0.1g or less/100g) | More than 1.5g/100g (More than 0.6g/100g) |

**Guideline amounts**
UK labels often indicate how low or high a food is in key nutrients based on government guidelines (above). Sometimes a colour-coded labelling system is used.

# WHAT IS A DIET?

Although many people associate the word "diet" with a calorie-controlled regimen designed to encourage weight loss, it has a much broader meaning. The true definition is the sum of the food and drink a person habitually eats. It can refer to a specific regimen followed for health reasons, such as a gluten-free diet, or an eating pattern that is based on personal preference or beliefs, such as a vegan diet. It can also be used to describe a way of eating that is based on the food habits in certain parts of the world (e.g. the Japanese or Mediterranean diet).

**NUTRITION**

To promote health and wellbeing

**GEOGRAPHY**

Such as Japanese, Mediterranean, and Nordic diets (see pp.122–123)

**BELIEFS**

For cultural or ethical reasons

**HEALTH**

For medical treatment or prevention

**WEIGHT MANAGEMENT**

To achieve or maintain a healthy weight

# THE HEALTHIEST FOOD TRADITIONS

The traditional diets of Japan, Mediterranean countries, and Nordic countries are associated with longevity and good health in later life – Japan has the highest life expectancy in the world. All three diets are also linked with a lower risk of cardiovascular disease, dementia, and other chronic diseases that are connected to aging. While lifestyle factors play a part and many foods in these diets differ, a common theme is the focus on seasonal and whole foods, seafood, fruit, vegetables, and small amounts of meat and dairy.

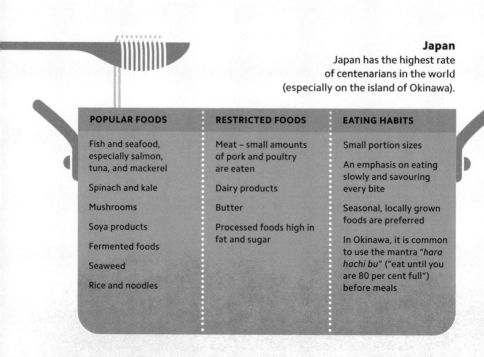

**Japan**
Japan has the highest rate of centenarians in the world (especially on the island of Okinawa).

| POPULAR FOODS | RESTRICTED FOODS | EATING HABITS |
|---|---|---|
| Fish and seafood, especially salmon, tuna, and mackerel | Meat – small amounts of pork and poultry are eaten | Small portion sizes |
| Spinach and kale | Dairy products | An emphasis on eating slowly and savouring every bite |
| Mushrooms | Butter | Seasonal, locally grown foods are preferred |
| Soya products | Processed foods high in fat and sugar | In Okinawa, it is common to use the mantra "*hara hachi bu*" ("eat until you are 80 per cent full") before meals |
| Fermented foods | | |
| Seaweed | | |
| Rice and noodles | | |

## Mediterranean countries

There is no one Mediterranean diet, but Mediterranean diets (e.g. in Greece, Italy, and Turkey) share some traits.

| POPULAR FOODS | RESTRICTED FOODS | EATING HABITS |
|---|---|---|
| Seasonal fruit and vegetables, grown locally | Red meat | Home cooking is common |
| Wholegrains | Butter | Meals are eaten together as a family |
| Beans and pulses | Processed foods high in fat and sugar | On-the-go snacking is rare |
| Oily fish and shellfish | | |
| Nuts and seeds | | |
| Fresh herbs | | |
| Garlic | | |
| Olive oil | | |

## Nordic countries

Studies have linked the diets of countries such as Sweden, Finland, Norway, and Denmark to longer life expectancy.

| POPULAR FOODS | RESTRICTED FOODS | EATING HABITS |
|---|---|---|
| Wholegrains (especially rye, oats, and barley) | Red meat, except for small amounts of game meats | Families tend to eat together |
| Fruit (primarily berries) and vegetables (mainly root vegetables) | Processed foods high in fat and sugar | On-the-go snacking is rare |
| Nuts | | An emphasis on seasonal foods, grown locally and organically |
| Oily fish such as herring, mackerel, and salmon | | |
| Rapeseed oil | | |

**Greener futures**

In 2016, a University of Oxford study estimated that a global switch to eating less meat and more vegetables would have huge benefits in fighting climate change and early deaths by 2050. The impact of a switch to a vegetarian diet would be even greater, and a vegan diet, greater still.

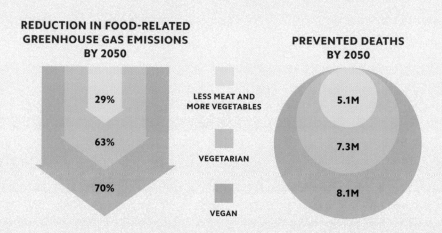

**REDUCTION IN FOOD-RELATED GREENHOUSE GAS EMISSIONS BY 2050**

29%

63%

70%

LESS MEAT AND MORE VEGETABLES

VEGETARIAN

VEGAN

**PREVENTED DEATHS BY 2050**

5.1M

7.3M

8.1M

# GOING GREEN

The current move towards plant-based diets is driven by mounting evidence that eating more plants and fewer animal products (especially red meat) is better for our health and the planet. Like the Oxford University research (see above), a 2019 study by the Harvard School of Public Health, using data from more than 81,000 participants, found that those who ate large amounts of meat were at the greatest risk of early death. Rearing animals for meat also damages the planet, contributing to deforestation and greenhouse gas (GHG) emissions (see opposite). Vegan and vegetarian diets are the most sustainable and, if widely adopted, could substantially reduce food-related GHG emissions, but simply eating less meat is beneficial, too.

# EATING FOR THE PLANET

Diet affects both an individual's health and the health of the planet. Food production accounts for up to 30 per cent of greenhouse gas (GHG) emissions and 70 per cent of fresh water usage. Agriculture is also responsible for deforestation, soil degradation, water pollution, and loss of biodiversity. Sustainable eating (see below) means choosing foods that have a low impact on the planet and its resources. It involves thinking about where food originates, and shopping and cooking habits, in order to minimize waste that goes to landfill, where bacteria produce methane (a greenhouse gas) as they break it down.

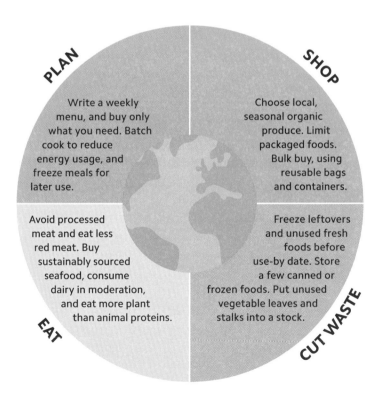

**PLAN**
Write a weekly menu, and buy only what you need. Batch cook to reduce energy usage, and freeze meals for later use.

**SHOP**
Choose local, seasonal organic produce. Limit packaged foods. Bulk buy, using reusable bags and containers.

**EAT**
Avoid processed meat and eat less red meat. Buy sustainably sourced seafood, consume dairy in moderation, and eat more plant than animal proteins.

**CUT WASTE**
Freeze leftovers and unused fresh foods before use-by date. Store a few canned or frozen foods. Put unused vegetable leaves and stalks into a stock.

| | VEGETABLES | FORTIFIED CEREALS | SOYA AND MYCOPROTEIN | PULSES | NUTS AND SEEDS |
|---|---|---|---|---|---|
| PROTEIN | | | All | All | All |
| OMEGA-3 FATS | | | | | Walnuts, chia, hemp, and flax seeds |
| CALCIUM | Leafy greens | Some (check labels) | Soya | All | Sesame seeds, almonds, Brazil nuts, and hazelnuts |
| IRON | Leafy greens | All | Tofu | All | All |
| ZINC | | | Mycoprotein | All | All |
| SELENIUM | Mushrooms | | | Chickpeas | Brazil nuts |
| IODINE | Seaweed | | | | |
| VITAMIN D | UV-exposed mushrooms | Some (check labels) | | | |
| VITAMIN B12 | | All | | | |

# PLANT-BASED PLANNING

Plant-based diets offer a number of well-documented health benefits (see p.124) but also require planning to ensure they provide all the nutrients necessary for good health and wellbeing. This is especially true for children, for whom growth processes have a high energy requirement, and people with increased nutritional needs, such as those who are breastfeeding or pregnant. Some nutrients, such as vitamins D and B12, and iodine, can be difficult to get enough of from plant-based foods alone. In these instances, supplements (see p.83) or fortified foods, such as breakfast cereals and plant-based dairy alternatives, are recommended.

# WHO EATS WHAT?

Plant-based diets vary in the foods they include, although all allow for fruit and vegetables (see below). Flexitarians do not remove any food group entirely but opt to eat less meat and more plants overall. Pescatarians may consume anything except meat. Lacto-ovo-vegetarians do not eat meat or fish. Ovo-vegetarians do not consume meat, fish, or milk products. Vegans do not consume any animal products at all (including honey). The plant-based diet an individual chooses might depend on one or many factors, including health concerns, or attitudes to animal welfare or climate change.

|  | 🥕 | 🥛 | 🥚 | 🐟 | 🥩 | 🍯 |
|---|---|---|---|---|---|---|
| **FLEXITARIAN** | ✓ | ✓ | ✓ | ✓ | ✓ | ✓ |
| **PESCATARIAN** | ✓ | ✓ | ✓ | ✓ |  | ✓ |
| **LACTO-OVO-VEGETARIAN** | ✓ | ✓ | ✓ |  |  | ✓ |
| **OVO-VEGETARIAN** | ✓ |  | ✓ |  |  | ✓ |
| **VEGAN** | ✓ |  |  |  |  |  |

# LOSING WEIGHT

For those who wish to lose weight, the sheer volume of diet plans can make it difficult to know which one to choose. The key to success for most people is finding a plan that is effective and suits them. There are three important considerations. Does the diet work? Is it healthy? Could it be followed long term? All such plans aim to create a calorie deficit, so the body has to use stored fat for energy, causing weight loss (see also pp.12–13). Eight popular diets are assessed below.

**WHAT IS IT?**  **CALORIE COUNTING**  **DOES IT WORK?**

Calories in food and drink are calculated to help ensure that a person adheres to a set daily calorie allowance.

This approach can be successful, especially if combined with exercise, but constantly counting calories is difficult to continue long term.

**LOW CARB/HIGH PROTEIN**

The daily intake of carbohydrates is restricted. Protein foods are unlimited and help reduce hunger as they are digested more slowly.

This can work short term, but low carbohydrate intakes make it hard to meet fibre needs. Protein foods may also be high in saturated fat.

**DIET CLUBS**

Diet clubs typically recommend reduced-calorie plans based on specially designed scoring systems that limit certain foods.

Group support can keep people motivated for longer, but group weigh-ins may not suit some, and membership can be pricey.

**KETO**

This very low-carbohydrate, high-fat diet encourages the body to break down fat, creating ketones, which are then used to provide energy.

It can produce short-term weight loss, but its very low carbohydrate content (20–50g daily) may increase the risk of chronic disease over time.

> Combining dieting with exercise can accelerate weight loss; even at rest, muscles burn much more energy than fat.

| WHAT IS IT? | LOW FAT | DOES IT WORK? |
|---|---|---|
| Gram for gram, fat is twice as energy dense as protein or carbohydrate, so a low-fat diet will generally reduce calorie intake. |  | These diets can be effective because they restrict calories, but many people find them hard to adhere to long term. |

| | TIME-RESTRICTED EATING | |
|---|---|---|
| The 5:2 diet limits calorie intake on two days a week to around 2,520kJ (600kcal). On the 16:8 plan, people eat within an 8-hour period each day. |  | Some research has shown that these diets are safe and have a positive effect on health, but sticking to them requires a lot of discipline. |

| | PALEO DIET | |
|---|---|---|
| This Stone-Age inspired diet includes meat, fish, nuts, seeds, fruit, and vegetables but excludes dairy, grains, sugar, and all other processed foods. |  | This can work, probably because it excludes several energy-dense food groups, but omitting them could lead to nutrient deficiencies. |

| | MEAL REPLACEMENT | |
|---|---|---|
| Formulated diet shakes, ready meals, and bars replace normal meals and snacks. Some plans allow one daily regular, calorie-controlled meal. |  | These work short term by cutting calories, but such diets can be monotonous and make family meals or eating out difficult. |

# SPECIAL DIETS

Some medical conditions require people to adopt special diets (see below). Often, this is because a person's symptoms are worsened or their health is threatened by certain types of food or drink so they must be avoided or eliminated entirely. It is essential to consult a registered dietician or other health professional who can advise on whether a special diet is appropriate and how to ensure it is nutritionally balanced.

| TYPE OF SPECIAL DIET | REASONS FOR SPECIAL DIET |
| --- | --- |
| Reduced-energy or low-calorie diets | To reduce body weight for those who are overweight (see p.134) |
| Exclusion and elimination diets (e.g. gluten-free, dairy-free, or lactose-free) | To manage coeliac disease (see p.146) or food allergies (see p.148) |
| Diets with increased energy or protein | To treat those with increased dietary needs – for example, after sustaining burns or undergoing surgery |
| Low-protein, low-sodium, low-potassium, or low-phosphorus diets | To manage kidney or liver problems |
| Controlled or low-carb diets | To control blood glucose levels in people with pre-diabetes, type 2 diabetes (see p.135), or a dependence on insulin |
| Modified texture diets, including soft foods or tube-feeding | To manage dysphagia (swallowing difficulties) or a sore mouth |
| High-fibre, low-fibre, or FODMAP diets | To manage digestive disorders such as irritable bowel syndrome (IBS) (see p.147) |
| Cholesterol-lowering diets | To reduce risk factors associated with cardiovascular disease |
| The DASH (Dietary Approaches to Stop Hypertension) diet | To reduce high blood pressure (see p.137) |

# TOO GOOD TO BE TRUE?

A diet may be a fad if it does any of the following:

 Promotes rapid weight loss – slowly and steadily is the safest way to lose weight and the best way to keep it off

 Demonizes certain foods – unless a person's diet is restricted by a medical condition (see opposite), there is no reason to fully eliminate any food

 Promotes miracle foods – there are no magic foods or drinks that will result in weight loss

 Offers a quick fix – products that promote weight loss without diet or lifestyle changes should be avoided

 Relies on testimonials and case studies but has no scientific research to support the claims it makes

 Links to supplements or other products – be especially wary of diets that push expensive products

There are many diet plans that offer advice on how to lose weight. Some are based on research but others – known as fad diets – spring up to push a product or cash in on a trend. Many promise quick fixes and rapid results, but may be more likely to result in nutritional deficiencies. They can also be difficult to sustain – weight lost rapidly may be quickly regained, and this can lead to an unhealthy cycle known as yo-yo dieting. The only way to lose weight and keep it off is to make long-term dietary and lifestyle changes.

# DIET
## AND HE
## CONDI

ALTH
TIONS

**Eating well can offer protection** against disorders such as heart disease, type 2 diabetes, tooth decay, obesity, and many types of cancer. Making good food choices can even help to reverse some conditions, such as high blood cholesterol and constipation, or improve symptoms in common complaints, like indigestion or insomnia. Certain conditions (e.g. anaemia) may require more specific dietary advice and, where food is responsible for health problems (e.g. coeliac disease or food allergies), diet modification is central to treatment.

# STORING TOO MUCH FAT

When a person's energy intake regularly exceeds their energy output (see pp.12–13), they can become overweight or obese. Excess body weight increases the risk of many chronic health conditions (e.g. type 2 diabetes, stroke). The body mass index (BMI) is a tool that can indicate if a person is a healthy weight based on their height. However, too much body fat (especially around the waist), not weight, is what increases health risk (see below). Studies show that effective long-term weight loss requires a combined strategy – a lower-energy diet with regular physical activity. Overweight and obesity are complex and individual, so always consult a health professional.

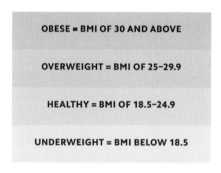

OBESE = BMI OF 30 AND ABOVE

OVERWEIGHT = BMI OF 25–29.9

HEALTHY = BMI OF 18.5–24.9

UNDERWEIGHT = BMI BELOW 18.5

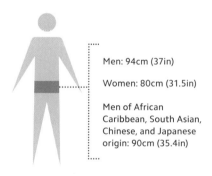

Men: 94cm (37in)

Women: 80cm (31.5in)

Men of African Caribbean, South Asian, Chinese, and Japanese origin: 90cm (35.4in)

**BMI**
The BMI measures weight, not body composition, so is an inaccurate measurement for some people. For example, athletes with a lot of muscle may have a high BMI despite having little body fat.

**Waist measurement**
Maximum advisable waist sizes vary by sex and ethnicity. Some men (see above) are more at risk of cardiovascular disease and diabetes with lower body fat.

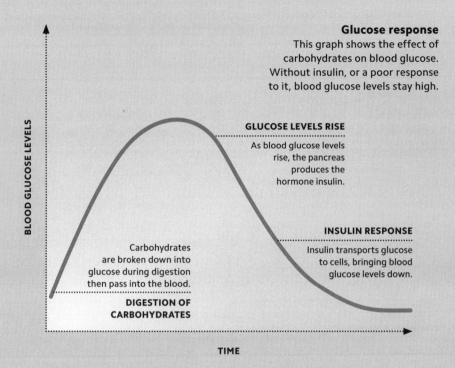

**Glucose response**
This graph shows the effect of carbohydrates on blood glucose. Without insulin, or a poor response to it, blood glucose levels stay high.

BLOOD GLUCOSE LEVELS

**GLUCOSE LEVELS RISE**
As blood glucose levels rise, the pancreas produces the hormone insulin.

**INSULIN RESPONSE**
Insulin transports glucose to cells, bringing blood glucose levels down.

Carbohydrates are broken down into glucose during digestion then pass into the blood.

**DIGESTION OF CARBOHYDRATES**

TIME

# ALL ABOUT INSULIN

Diabetes is a condition characterized by elevated blood glucose levels. If left untreated or poorly controlled, it can severely damage the heart, eyes, kidneys, brain, and nerves. Globally, it affects one in 10 adults. Type 1 diabetes (an autoimmune condition) is when the pancreas stops producing insulin. This type is treated with insulin injections. With type 2 diabetes, the body fails to make enough insulin or does not fully respond to it (insulin resistance), so blood glucose stays high (hyperglycaemia). This type makes up 90 per cent of cases and weight loss is part of prevention, along with a healthy diet that limits free sugars.

# CONTROLLING CHOLESTEROL

Diet changes can help reduce high blood cholesterol. The first step is to replace saturated fats, which increase "bad" (LDL) cholesterol, with monounsaturated and polyunsaturated fats. These help maintain "good" (HDL) cholesterol, which actually removes LDL cholesterol from blood vessels (see below). Other diet components that help reduce LDL cholesterol include beta-glucan, a fibre in oats and barley (see pp.20–21), soya protein, nuts and seeds, and products fortified with plant stanols and sterols. Antioxidant-rich foods (e.g. fruit and vegetables) can also help reduce oxidation of LDL cholesterol, which triggers inflammation in the artery walls, a risk factor for hardening of the arteries (atherosclerosis).

**A HEALTHY ARTERY**

A blood vessel without cholesterol build up means blood can flow easily through the arteries.

**CHOLESTEROL BUILD UP**

Saturated fats build up and create fatty plaques that narrow and harden the arteries.

Blood clot blocks artery

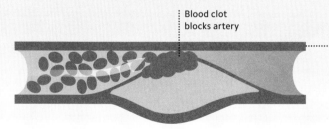

**BLOOD CLOTS**

When fatty plaques burst or break apart, blood clots can form and block an artery, risking a heart attack.

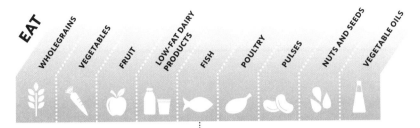

**EAT**
WHOLEGRAINS · VEGETABLES · FRUIT · LOW-FAT DAIRY PRODUCTS · FISH · POULTRY · PULSES · NUTS AND SEEDS · VEGETABLE OILS

Omega-3 fats contribute to normal blood pressure, so eat two portions of fish a week (one oily).

**LIMIT**
FATTY MEAT · FULL-FAT DAIRY PRODUCTS · HIGH-SUGAR FOODS AND ADDED SUGAR · HIGH-SALT FOODS AND ADDED SALT · COCONUT AND PALM OIL · ALCOHOL · SUGARY DRINKS

**The DASH diet**
The DASH (Dietary Approaches to Stop Hypertension) diet recommends eating particular foods and limiting others.

Check labels for salt content. Foods with more than 1.5g salt/100g or 1.8g salt per serving are high in salt.

# LOWERING BLOOD PRESSURE

Hypertension (high blood pressure) is when the pressure in blood vessels is too high. Hypertension is sometimes called the silent killer as it often shows no symptoms, yet increases the risk of cardiovascular disease. A high intake of sodium – a key component of salt – is a risk factor (see p.75) for high blood pressure, so the first line of treatment is typically to reduce salt intake by not adding salt to food and eating fewer salty foods. Other risk factors include being overweight (see p.134) and drinking too much alcohol (see p.80).

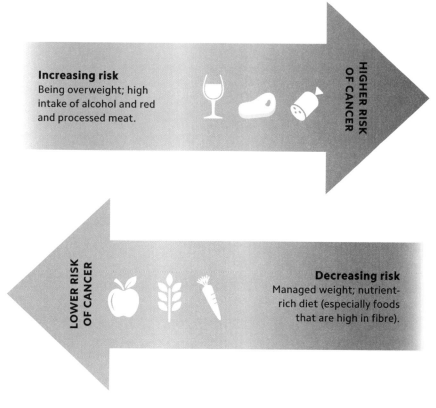

**Increasing risk**
Being overweight; high intake of alcohol and red and processed meat.

HIGHER RISK OF CANCER

LOWER RISK OF CANCER

**Decreasing risk**
Managed weight; nutrient-rich diet (especially foods that are high in fibre).

# CUTTING CANCER RISK

There is no one diet, food, or nutrient that can prevent or treat cancer, but eating healthily offers protection. Managing weight is key as being overweight is the biggest cause of cancer after smoking and is linked to 13 different types. The fibre and antioxidants in nutrient-rich plant foods generally offer the most protection, and these foods (low in saturated fat, sugar, and salt) are also linked to better weight management. For those living with cancer, there is no evidence that alternative diets, which claim to cure the disease, are effective. Often, they restrict nutritious foods and can do more harm than good.

| CAUSES | Various factors can cause malnutrition. While often associated with low weight, it occurs in people who are overweight, too. | **ILLNESS OR INFECTION** |
| | | **LOW MOOD OR IMPAIRED MENTAL HEALTH** |
| | | **PHYSICAL OR MENTAL DISABILITY** |
| | | **POOR APPETITE** |
| | | **DIFFICULTY CHEWING OR SWALLOWING** |
| | | **POVERTY** |
| | | **CHANGES TO SMELL OR TASTE** |

# UNDERNOURISHED

Malnutrition occurs when the body lacks the energy and/or nutrients it needs to function. Although more common in older people and people with low body weight, it can result at any age from various physical and psychological factors. Malnutrition increases the body's vulnerability to disease, impairs wound healing, reduces fertility, causes deficiency disorders (e.g. anaemia), stunts growth, and results in a loss of muscle mass and strength. Seeking professional help to identify the causes of malnutrition and to treat it effectively is essential.

| MANAGEMENT | Diagnosis and treatment is an essential first step, accompanied by social help, if necessary, and an appropriate nutrient-rich diet. | **MEDICAL ADVICE** |
| | | **PSYCHOLOGICAL SUPPORT** |
| | | **SOCIAL SERVICES ADVICE (IF NECESSARY)** |
| | | **A NUTRIENT-RICH, BALANCED DIET** |
| | | **ENERGY-RICH FOODS** |
| | | **SMALL, FREQUENT MEALS** |

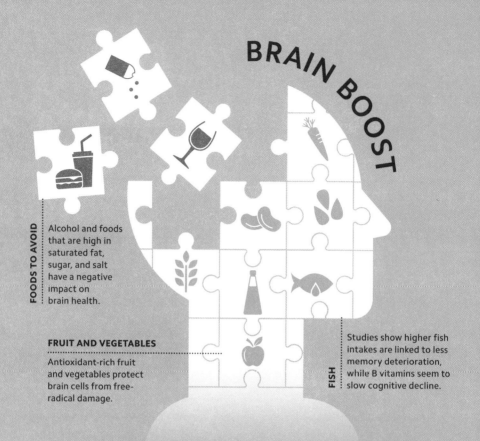

# BRAIN BOOST

(see pp.122–123)

(see p.130)

**FOODS TO AVOID**

Alcohol and foods that are high in saturated fat, sugar, and salt have a negative impact on brain health.

**FRUIT AND VEGETABLES**

Antioxidant-rich fruit and vegetables protect brain cells from free-radical damage.

**FISH**

Studies show higher fish intakes are linked to less memory deterioration, while B vitamins seem to slow cognitive decline.

Dementia refers to the symptoms caused by diseases that damage the brain (e.g. Alzheimer's disease and vascular dementia). There is a strong link with heart disease – both have similar risk factors such as high cholesterol and high blood pressure. To help protect against dementia, the MIND (Mediterranean-DASH Intervention for Neurodegenerative Delay) diet combines the heart-healthy Mediterranean diet (see pp.122–123) with the DASH diet (see p.130), which lowers blood pressure. This eating plan is low in saturated fat, sugar, and salt, and rich in seafood and plant-based foods.

# FEEDING THE MIND

Dietary choices not only impact physical health but mental health, too. A nutritious diet can help ease symptoms linked to common mental health problems such as stress, anxiety, and depression. Eating regularly and including fibre-filled starchy carbohydrates (see p.66) regulates blood glucose levels, avoiding dips that may compound symptoms such as a lack of energy or irritability. Some protein-rich foods contain tryptophan, an amino acid that makes the chemical serotonin, which boosts feelings of happiness. Studies show that diets rich in fish, beans, wholegrains, fruit, and vegetables – such as the Mediterranean diet (see pp.122–123) – help protect against depression.

| HEALTHY DIET | UNHEALTHY DIET |
| --- | --- |
| EATING REGULARLY | EXCESSIVE ALCOHOL INTAKE |
| B VITAMINS | EXCESSIVE CAFFEINE INTAKE |
| FRUIT AND VEGETABLES | SKIPPING MEALS |
| OMEGA-3 FATS/OILY FISH | DEHYDRATION |

**Positive choices**
Healthy dietary choices (see pp.62–63) can promote positive mental health as they boost a sense of wellbeing.

**Negative choices**
Caffeine can increase anxiety. Low mood can be caused by too much alcohol (see p.80), skipping meals (see p.115), or dehydration.

# DISORDERED HABITS

Eating disorders are serious, complex mental illnesses that can affect anyone, regardless of gender, age, or ethnicity. Disordered eating behaviours – which include limiting food, eating very large amounts in one go, and using unhealthy methods to eliminate eaten food (e.g. through vomiting) – are used to cope with difficult feelings. There are many types of eating disorders, including anorexia nervosa, bulimia nervosa, and binge eating disorder (BED). If an eating disorder fails to match specific diagnostic criteria, a diagnosis of "other specified feeding or eating disorder" (OSFED) may be made. All eating disorders can seriously harm physical and mental health, but full recovery is possible by seeking professional help.

# PROTECTING TEETH

Dental decay and gum disease are widespread problems but are preventable with good dental hygiene and a diet that limits sugar and acidic foods and drinks (see below). Bacteria in dental plaque feed on carbohydrate food residue, creating acids that damage tooth enamel. All carbohydrate foods can potentially cause damage, but sugars have the greatest effect. Acidic foods, fruit juices, and fizzy drinks also dissolve tooth enamel (dental erosion). When teeth are frequently exposed to acids, a cavity can form, leading to tooth decay.

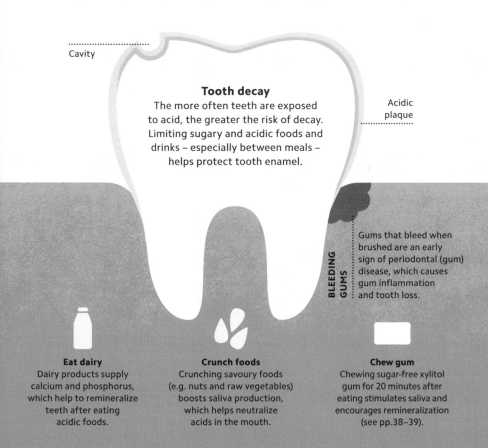

Cavity

**Tooth decay**
The more often teeth are exposed to acid, the greater the risk of decay. Limiting sugary and acidic foods and drinks – especially between meals – helps protect tooth enamel.

Acidic plaque

BLEEDING GUMS
Gums that bleed when brushed are an early sign of periodontal (gum) disease, which causes gum inflammation and tooth loss.

**Eat dairy**
Dairy products supply calcium and phosphorus, which help to remineralize teeth after eating acidic foods.

**Crunch foods**
Crunching savoury foods (e.g. nuts and raw vegetables) boosts saliva production, which helps neutralize acids in the mouth.

**Chew gum**
Chewing sugar-free xylitol gum for 20 minutes after eating stimulates saliva and encourages remineralization (see pp.38–39).

**Factors outside the stomach**
Overeating, being overweight or obese, some medical problems (e.g. hiatus hernia), some medication, pregnancy, stress and anxiety, and smoking.

**Factors inside the stomach**
Alcohol, caffeine, carbonated drinks, fatty foods, spicy foods, and acidic foods.

**Acid reflux**
Many factors increase the risk of acid reflux (see above). It is better to address these than to rely on over-the-counter antacids to help relieve symptoms.

# EASING THE STOMACH

Indigestion (dyspepsia) can cause symptoms that include heartburn, nausea, flatulence, and belching. It is caused by acid reflux, where stomach acids travel into the oesophagus towards the throat, causing an unpleasant, bitter taste in the mouth. If acid reflux keeps happening, it is called gastro-oesophageal reflux disease (GORD), which can result in other symptoms, such as a recurring cough or hiccups, a hoarse voice, and bad breath.

# AIDING DIGESTION

Constipation can affect anyone, including babies and children, but is more common in women, older adults, and during pregnancy. Symptoms include hard stools, infrequent bowel movements, stomach pain, bloating, flatulence, poor appetite, and nausea. Constipation is often caused by a diet that lacks sufficient fibre and fluid. However, inactivity, stress, ignoring the urge to defecate, and some medications (e.g. iron supplements) and conditions, such as IBS (see p.147), increase the risk. Untreated, constipation can become chronic and cause haemorrhoids (piles). Increasing fluid intake and eating more fibre-rich foods (see pp.20–21) will help produce softer, bulkier stools that are easier to pass. Fibre should be increased gradually to avoid bloating and cramping.

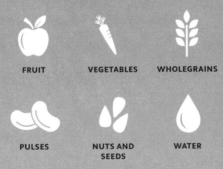

FRUIT     VEGETABLES     WHOLEGRAINS

PULSES     NUTS AND SEEDS     WATER

**First line of treatment**
For many who experience constipation, the issue is dietary. Staying hydrated (see p.43) and eating fibre-rich foods should help symptoms to subside.

# AVOIDING GLUTEN

Coeliac disease is not an allergy but an autoimmune disease, where the body's immune system reacts abnormally to gluten (see below). Symptoms can vary in severity but often include stomach pain and digestive problems, tiredness, mouth ulcers, weight loss, anaemia, and a skin rash (dermatitis herpetiformis). There is no cure or medication, but a lifelong diet that eliminates all gluten-containing foods – such as pasta, couscous, many breakfast cereals, bread, cake, biscuits, pizza bases, and pastry products made from wheat, barley, or rye – can manage the condition, preventing symptoms and allowing the gut to heal.

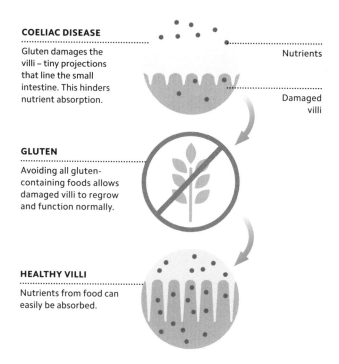

**COELIAC DISEASE**

Gluten damages the villi – tiny projections that line the small intestine. This hinders nutrient absorption.

Nutrients

Damaged villi

**GLUTEN**

Avoiding all gluten-containing foods allows damaged villi to regrow and function normally.

**HEALTHY VILLI**

Nutrients from food can easily be absorbed.

# SOOTHING THE GUT

Irritable bowel syndrome (IBS) describes a range of digestive symptoms that can include stomach pain or cramps, bloating, diarrhoea, and constipation. The cause of IBS is unknown, but it is thought to be linked to stress, oversensitive gut nerves, and food passing through the digestive system too quickly or slowly. Although there is no cure, dietary strategies (see right) can help. However, if symptoms persist, a diagnosis from a doctor is important to rule out other conditions, such as coeliac disease (see opposite) and inflammatory bowel disease. Professional help can also advise on other forms of treatment.

**EAT HEALTHILY**

Do not skip meals; eat regularly and slowly; limit fizzy, alcoholic, and caffeinated drinks; eat fewer fatty and spicy foods; limit fruit to three daily portions; consider trying a probiotic supplement for at least four weeks.

**TRACK SYMPTOMS**

Keep a food, mood, and symptoms diary to identify links between symptoms and foods, events, or emotions. Some foods, for example, are gas-producing (e.g. beans, onions), while too many high-fibre foods may worsen diarrhoea.

**MODIFY DIET**

If the food, mood, and symptoms diary links a food with symptoms, eliminate it. If symptoms improve, reintroduce the food and see if symptoms recur. Seek advice from a dietitian if many foods seem to be problematic.

**IF SYMPTOMS PERSIST**

A health professional may recommend a low-FODMAP diet which limits the intake of carbohydrates found in fruit, vegetables, milk, and pulses that are not easily broken down in the gut.

# ALLERGIC REACTIONS

A food allergy is an abnormal immune response to a specific food. It can be immediate or delayed. With an immediate food allergy, the body wrongly identifies a protein in a food (e.g. nuts) as harmful and rushes to protect itself by producing immunoglobulin E (IgE) antibodies. This releases chemicals (e.g. histamine) that trigger symptoms ranging from mild to life-threatening (anaphylaxis). A delayed food allergy involves the immune system but is not mediated by IgE, so reactions vary, and symptoms can occur days later.

### Incidence
Around 5–8 per cent of young children and 1–2 per cent of adults have a food allergy.

### Cause
A food allergy is caused by an immune response to a protein in food. This produces antibodies.

### Body responses
Immediate reactions occur in minutes or within two hours; delayed reactions can take days.

### Symptoms
Food allergy symptoms include vomiting, itching, swelling of the lips and eyes, and anaphylaxis.

### Diagnosis
To diagnose an allergy, a health professional must oversee blood and skin prick tests.

**COMMON ALLERGIES**

| MILK | EGGS | PEANUTS | TREE NUTS | WHEAT | SHELLFISH |

# NON-ALLERGIC REACTIONS

A food intolerance does not involve the immune system. It occurs when the body is sensitive to an ingredient, such as salicylates (e.g. in some fruit, vegetables, and nuts) or histamine (e.g. in salami), or is unable to effectively process it, such as lactose in milk. Food intolerances can cause a vast range of immediate or delayed reactions that vary in severity. As the immune system is not involved, it is harder to diagnose food intolerances. Sometimes, small amounts of food can also be tolerated without side effects.

**Incidence**
Studies suggest that 15–20 per cent of people worldwide have a food intolerance.

**Cause**
A food intolerance occurs when the body is sensitive to a food or cannot process it.

**Body responses**
Reactions prompted by a food intolerance can be quick but often develop more slowly.

0–72 HRS

**Symptoms**
The effects of a food intolerance include rashes, eczema, stomach pain, and digestive issues.

**Diagnosis**
To diagnose an intolerance, a registered dietitian can oversee a food elimination and reintroduction diet.

**COMMON INTOLERANCES**

LACTOSE

CAFFEINE

GLUTEN

ALCOHOL

SALICYLATES

HISTAMINE

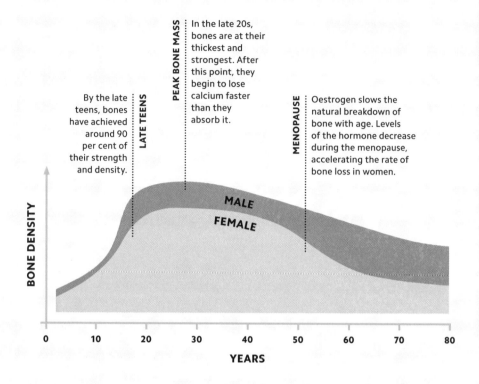

By the late teens, bones have achieved around 90 per cent of their strength and density.

**LATE TEENS**

**PEAK BONE MASS**

In the late 20s, bones are at their thickest and strongest. After this point, they begin to lose calcium faster than they absorb it.

**MENOPAUSE**

Oestrogen slows the natural breakdown of bone with age. Levels of the hormone decrease during the menopause, accelerating the rate of bone loss in women.

MALE

FEMALE

BONE DENSITY

| 0 | 10 | 20 | 30 | 40 | 50 | 60 | 70 | 80 |

YEARS

# PREVENTATIVE MEASURES

Osteoporosis is a condition where the bones become so weak that they are prone to breaking. Worldwide, one in three women and one in five men over 50 will suffer a fracture due to osteoporosis. Prevention begins in childhood. A bone-friendly diet (see p.108), especially in adolescence, creates a higher peak bone mass, so bones remain stronger. Limiting caffeine and alcohol, maintaining a healthy weight, and not smoking also offer protection. Older men and women (see above) who are at risk of or diagnosed with osteoporosis may be prescribed calcium and vitamin D supplements and should avoid high retinol (vitamin A) intakes, which can weaken bones.

# EASING JOINT PAIN

There are many different types of arthritis. The two most
common are osteoarthritis and rheumatoid arthritis (see below),
but all cause painful inflammation in joints. There is no cure but
there are treatments that can slow its progression. Weight loss,
if necessary, can reduce pressure on joints and a Mediterranean
diet (see pp.122–123), rich in anti-inflammatory nutrients, may
help relieve symptoms. Studies also suggest a link between the
gut microbiome and rheumatoid arthritis, providing more
evidence for eating prebiotic and probiotic foods (see p.106).

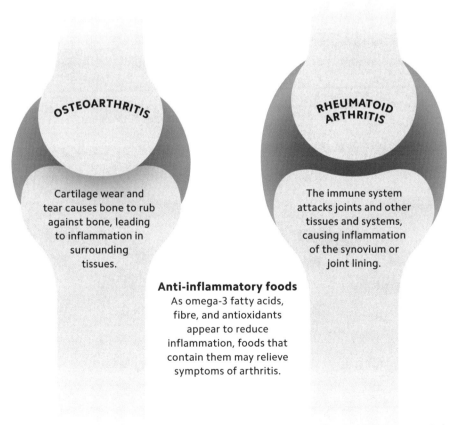

**OSTEOARTHRITIS**

Cartilage wear and
tear causes bone to rub
against bone, leading
to inflammation in
surrounding
tissues.

**RHEUMATOID ARTHRITIS**

The immune system
attacks joints and other
tissues and systems,
causing inflammation
of the synovium or
joint lining.

**Anti-inflammatory foods**
As omega-3 fatty acids,
fibre, and antioxidants
appear to reduce
inflammation, foods that
contain them may relieve
symptoms of arthritis.

# BLOOD WORK

There are many types of anaemia, but the most common is iron-deficiency anaemia, which affects around 500 million people worldwide. It is caused by low iron levels, resulting in fewer red blood cells than usual. This means that there is less capacity for the blood to carry oxygen around the body. Symptoms of iron-deficiency anaemia include pale skin, extreme fatigue, heart palpitations, and shortness of breath. The first line of treatment is to increase dietary consumption of iron (see p.34), but a doctor may also prescribe iron supplements. A lack of vitamins B9 and B12 (see pp.24–25) can also result in a form of anaemia where blood cells are abnormally large so do not function properly.

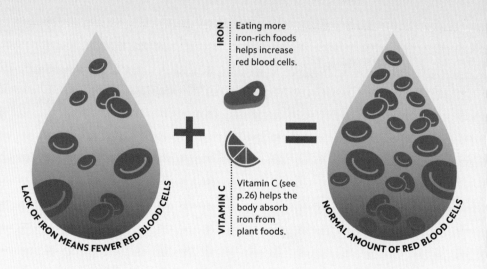

IRON
Eating more iron-rich foods helps increase red blood cells.

VITAMIN C
Vitamin C (see p.26) helps the body absorb iron from plant foods.

LACK OF IRON MEANS FEWER RED BLOOD CELLS

NORMAL AMOUNT OF RED BLOOD CELLS

### Eat nutrient-dense foods
For example, homemade soups, scrambled eggs, steamed fish, yogurt, and dairy beverages.

### Get enough fluid
Infections can cause fever, sweating, vomiting, and diarrhoea, increasing fluid loss and the risk of dehydration.

### Improve gut health
Antibiotics can disrupt the gut microbiome (see p.106), so taking specific probiotics during and for a week after antibiotic use may support gut health.

### Consider a multi-vitamin supplement
When food intake is reduced, multi-vitamins provide a safeguard to ensure adequate intakes.

# FIGHTING INFECTION

Many nutrients support immunity (see p.107), but no single food or nutrient can stop or treat infections such as colds, COVID-19, pneumonia, gastroenteritis, urinary infections, or infected wounds. However, when fighting an infection, energy and nutrient requirements increase so nutrient-dense foods (see p.61) can support healing, while malnutrition (see p.139) slows recovery. Eating can be challenging when unwell, so eating small amounts of nutrient-rich foods more frequently can help, along with drinking plenty of fluids. Rehydration salts, probiotics, and vitamin and mineral supplements can also support recovery.

# EATING FOR HORMONAL HARMONY

Premenstrual syndrome (PMS) affects many women in the second half of the menstrual cycle and is probably caused by changing hormone levels. Symptoms include mood swings, irritability, tiredness, breast tenderness, headaches, and skin problems (e.g. acne). More severe symptoms are known as premenstrual dysphoric disorder (PMDD). PMS can affect eating habits – many people get cravings, rely on sugary snacks to boost energy, or comfort eat. Eating regular meals that include the nutrients below may help ease symptoms.

**Fibre**
Eating fibre-rich foods (see pp.20–21) can help ease PMS symptoms by keeping blood glucose levels steady.

**Omega-3 fats**
Studies suggest these can help reduce the severity of PMS symptoms. Oily fish is the best source.

**Calcium and Vitamin D**
Research shows that low blood levels of these nutrients can cause or exacerbate PMS symptoms.

**B vitamins**
Vitamin B6 (see p.24) helps to regulate hormone activity.

**MILK**

.....................................

Milk is rich in tryptophan and calcium, low levels of which have been linked to sleep problems.

**HERBAL TEA**

.....................................

Some herbal teas can help induce sleep (e.g. chamomile, which is a natural sedative).

**CAFFEINE**

.....................................

Avoid caffeinated coffee, tea, green tea, cola, or energy drinks for at least four hours before bed.

**ALCOHOL**

.....................................

Drinking alcohol impairs sleep quality as the body works to process it.

# EATING FOR SLEEP

During sleep, the body performs essential restorative processes. Therefore, regular sleep deprivation can raise the risk of health problems, such as heart disease, high blood pressure, type 2 diabetes, weakened immunity, and depression. However, certain foods (see above) can support sleep. Protein-based foods are especially key, as most contain tryptophan, an amino acid that helps make serotonin, which is then converted into the sleep-inducing hormone melatonin. Stimulants (e.g. caffeine) should be avoided before bedtime, as well as large meals (or high-fat/spicy food), which can lead to indigestion.

# INDEX

Page numbers in **bold** refer to main entries.

molybdenum 22, **38–39**
monosaccharides 18
monounsaturated fatty
acids 16, 93, 111
mood 46, 141
mouth 47, **48**, **103**
mucilages 20
mucous membranes 23, 25,
103
muscles 12, 15, 19, 27, 30,
31, 33, 88, 89, 97, 109,
**110**, 139
mycoprotein 71, 72

# N

national dietary guidelines
**62–63**
natural sugars 74
nervous system 25, 26, 28,
30, 33, 37, 39, 100, 140
neurotransmitters 100
niacin 24–25
nitrates 104
non-essential amino acids 14
Nordic countries, diet 121,
**123**
nutrient density **61**, 153
nutrients
absorption **51**, 56
bioavailability **56–57**
food preparation/cooking
57
malabsorption **53**
nuts and seeds 71, 82, 136

# O

obesity 21, 77, 95, 116, **134**,
137
oesophagus 47, **48**, 144
oestrogen 150
oils 63, **73**
older adults 88, **97**
oligosaccharides 18, 20
omega-3 fats 91, 94, 100,
102, 104, 109, 111, 137,
151, 154

osteoarthritis 151
osteoporosis 27, 31, 108, **150**
ovo-vegetarians 127
oxygen 152

# P

paleo diet 129
pancreas 47, **50**, 135
pantothenic acid 24–25
pasteurizing 81
pectins 20
pepsin 49
peptides 50
pescatarians 127
phospholipids 17
phosphorus 22, **32**, 68
photosynthesis 18
phylloquinone 29
physical activity 13
phytonutrients 10, **40–41**,
64, 67
pickling 81
planet, health of the 124, 125
plant-based foods/diets 7,
31, 37, **40–41**, 55, 62,
**64–65**, **71**, **124**, **126–127**
plaque
arterial 136
dental 143
pollution 125
polyphenols 41, 104, 106
polysaccharides 18
polyunsaturated fatty acids
16
population levels 72
portion size **117**, 118
potassium 22, **30**, 65, 104,
110
prebiotics 20, 106, 151
pregnancy 93, **94**, 126
premenstrual syndrome
(PMS) **154**
preschool children **89**
preservation, food 76, **81**
probiotics 55, 106, 151
processing 57, 66, 67, 76, **77**,
93, 116, 140

prostaglandins 17
protein 10, 11, 12, 14, **15**, 61,
63, 68, 90, 101, 110, 117,
155
animal-based **70**
new **72**
plant **71**
pulses 71, 93
pyridoxine 24–25

# QR

quercetin 41
recommended daily
amounts (RDA) 92, 120
rectum 47, 52
red blood cells 25, 105, 152
resistance exercise 110
retinol 23, 108, 150
reward, food as 118
rheumatoid arthritis 151
riboflavin 24–25
roasting 83

# S

safety, food 120
salicylates 149
saliva 47, 48, 143
salt/salting **75**, 81, 137, 140
satiety 21, 60, 119
saturated fats 16, 62, 68, 71,
73, 77, 104, 128, 136, 140
scurvy 26
seasonal produce 122, 125
selenium 22, **36**, 70, 101, 107
serotonin 155
shopping habits 125
simple carbohydrates 18
skin 23, 25, 35, **111**
rashes 149
sleep problems 80, **155**
small intestine 20, 47, 50,
**51**, 106, 146
smoking 138, 150
snacks 90, 114, **116**, 118
soaking 57
social events 118

# ACKNOWLEDGMENTS

DK would like to thank the following for their help with this book: Vanessa Hamilton and Mark Lloyd for the illustrations; Ginger Hultin for consulting; Kristina Routh for fact checking; Bonnie Macleod for editorial assistance; Debra Wolter for proofreading; Helen Peters for the index; Senior Jacket Designer Suhita Dharamjit; Senior DTP Designer Harish Aggarwal; Senior Jackets Editorial Coordinator Priyanka Sharma.

**150 Shutterstock.com:** eranicle (background)

All images © Dorling Kindersley
For further information see:
www.dkimages.com

# SIMPLY EXPLAINED

ARTIFICIAL INTELLIGENCE

FOR COMPLETE BEGINNERS

ASTRONOMY

FOR COMPLETE BEGINNERS

THE BRAIN

FOR COMPLETE BEGINNERS

CLIMATE CHANGE

FOR COMPLETE BEGINNERS

ECONOMICS

FOR COMPLETE BEGINNERS

MATHS

FOR COMPLETE BEGINNERS

NUTRITION

FOR COMPLETE BEGINNERS

PHILOSOPHY

FOR COMPLETE BEGINNERS

PSYCHOLOGY

FOR COMPLETE BEGINNERS

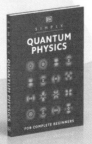

QUANTUM PHYSICS

FOR COMPLETE BEGINNERS

DK

For the curious